Model Predictive Control for Doubly-Fed Induction Generators and Three-Phase Power Converters

Model Predictive Control for Doubly-Fed Induction Generators and Three-Phase Power Converters

Alfeu J. Sguarezi Filho
Engineering, Modeling and Applied Social Sciences Center
Federal University of ABC
Santo André, SP, Brazil

ELSEVIER

Elsevier
Radarweg 29, PO Box 211, 1000 AE Amsterdam, Netherlands
The Boulevard, Langford Lane, Kidlington, Oxford OX5 1GB, United Kingdom
50 Hampshire Street, 5th Floor, Cambridge, MA 02139, United States

Library of Congress Cataloging-in-Publication Data
A catalog record for this book is available from the Library of Congress

British Library Cataloguing-in-Publication Data
A catalogue record for this book is available from the British Library

ISBN: 978-0-323-90964-8

Publisher: Joseph P. Hayton
Acquisitions Editor: Graham Nisbet
Editorial Project Manager: Andrae Akeh
Production Project Manager: Poulouse Joseph
Designer: Greg Harris

Typeset by VTeX

This book is dedicated to my beloved son Nicolas Sguarezi, my wife, and my family.

Contents

List of figures xiii
List of tables xvii
List of contributors xix
Biography xxi
Abbreviation list xxiii

1. Introduction
Alfeu J. Sguarezi Filho

1.1 Overview 1
1.2 Structure of the book 3

2. Induction machine and three-phase power converter dynamic models
Alfeu J. Sguarezi Filho

2.1 Space vector notation 5
2.1.1 The stationary reference frame ($\alpha\beta$) 5
2.1.2 The synchronous reference frame (dq) 6
2.2 Induction machine dynamic model 7
2.2.1 IM representation in three-phase systems 8
2.2.2 IM representation in stationary reference frame ($\alpha\beta$) 9
2.2.3 IM representation in synchronous reference frame (dq) 12
2.2.4 Speed dynamics representation 14
2.3 Three-phase power converter connected to the grid dynamic model 14
2.3.1 Three-phase power CCG representation in three-phase systems 15
2.3.2 CCG representation in stationary frame ($\alpha\beta$) 16
2.3.3 CCG representation in synchronous frame dq 17
2.4 Pulse-width-modulation techniques 18
2.4.1 Sinusoidal PWM 18
2.4.2 Space vector modulation 21
2.5 Summary 25
2.6 Further reading 25

3. Fundamentals of vector control for DFIG and for the three-phase CCG
Alfeu J. Sguarezi Filho

3.1 Doubly-fed induction generator 27
3.1.1 Vector control 28
3.1.2 Closed loop rotor current control using PI controllers 31
3.1.3 Deadbeat rotor current control for DFIG 33
3.1.4 Deadbeat direct power control for DFIG 35
3.2 Three-phase power CCG vector control 37
3.2.1 Filter elements 38
3.2.2 Vector control fundamentals for the CCG 38
3.3 Summary 40
3.4 Further reading 40

4. Fundamentals of model predictive control
Alfeu J. Sguarezi Filho

4.1 Overview 43
4.1.1 MPC applied in power electronics systems 45
4.2 Finite control set model predictive control 45
4.2.1 Principles of finite control set model predictive control 45
4.2.2 Constrains in FCS-MPC 47
4.2.3 Modulated finite control set model predictive control 47
4.3 MPC with modulator (MPC-WM) 48
4.3.1 Constrains in MPC 52
4.4 Summary 52
4.5 Further reading 53

5. Modulated FCS-MPC for DFIG-DPC
Alfeu J. Sguarezi Filho and Rogério V. Jacomini

5.1 Representation of DFIG using DPC 55
5.1.1 Rotor voltage representation 55
5.2 DPC for DFIG using the modulated FCS-MPC 57
5.2.1 Computation of the duty cycle using modulated FCS-MPC 58
5.3 Experimental results 63
5.4 Summary 66

6. A wireless coded modulated FCS-MPC DPC for renewable energy sources in smart grid environment
Alfeu J. Sguarezi Filho, Angelo S. Lunardi, Carlos E. Capovilla, and Ivan R.S. Casella

6.1 Overview 67
6.2 Three-phase power CCG using direct power predictive control 69
6.3 Representation of the wireless communication system 71

6.4 Analysis of the experimental results 75
6.4.1 OFDM-CC results 76
6.4.2 OFDM-LDPC results 77
6.4.3 Fast Fourier transform analysis 80
6.5 Summary 83

7. MPC-WM for doubly-fed induction generator and three-phase CCG

Alfeu J. Sguarezi Filho

7.1 DFIG rotor current control using MPC-WM 85
7.1.1 Space state equations 85
7.1.2 Rotor current control using MPC-WM 87
7.2 DFIG DPC using MPC-WM 91
7.2.1 DPC using MPC-WM 94
7.3 Three-phase CCG current control using MPC-WM 97
7.3.1 Space state equations 98
7.3.2 Grid current control using MPC-WM 100
7.3.3 Simulation results 101
7.3.4 Experimental results 101
7.4 Information about the choice of weighting matrices and horizons values 103
7.5 Summary 105

8. Fundamentals of the model predictive repetitive control

Alfeu J. Sguarezi Filho

8.1 Fundamentals of repetitive control 107
8.1.1 IMP for any periodic signal 107
8.1.2 Basic RC structure and design 108
8.2 Fundamentals of model predictive repetitive control 109
8.2.1 Periodic signals representation 109
8.2.2 MPRC technique 111
8.3 Summary 115

9. MPRC-WM for DFIG and three-phase CCG operation under voltage distortions

Alfeu J. Sguarezi Filho, Angelo S. Lunardi, and Eliomar R. Conde D.

9.1 Representation of voltage distortions 117
9.2 Model of DFIG under stator distorted voltage 118
9.2.1 Influence of distorted voltage in the stator active and reactive power representation 119
9.2.2 Influence of distorted voltage in DC link voltage 121
9.3 DFIG rotor current control using MPRC-WM 123

9.3.1 Criterion for choosing polynomial $D(z)$ 126
9.3.2 Experimental results 128
9.4 Three-phase power CCG model under grid distorted voltage 131
9.4.1 Influence of distorted voltage in the active and reactive power representation 133
9.5 Three-phase power CCG current control using MPRC-WM 134
9.5.1 Criterion for choosing the polynomial $D_g(z)$ 139
9.5.2 Experimental results 139
9.6 Summary 143

10. Finite position set phase-locked loop operating under nonideal grid voltages

Alfeu J. Sguarezi Filho, Fernando Lino, and Rogério V. Jacomini

10.1 PLL fundamentals 145
10.1.1 PLL for three-phase systems 146
10.2 Representation of grid voltage disturbances 147
10.3 Finite position set PLL operation under grid disturbances 149
10.3.1 Representation of the DSOGI 149
10.3.2 Representation of the MAF 151
10.3.3 Finite position set PLL 152
10.4 Experimental results 153
10.5 Summary 155

11. Implementation of DFIG MPC-WM and three-phase power CCG MPRC-WM using Simulink/MATLAB®

Alfeu J. Sguarezi Filho, Angelo S. Lunardi, and Eliomar R. Conde D.

11.1 Introduction 157
11.2 Building embedded functions for Park–Clarke transformation 157
11.2.1 Park–Clarke transformation 157
11.2.2 Inverse Park–Clarke transformation 159
11.2.3 Pulse width modulation 160
11.3 Building simulation model for DFIG 168
11.3.1 Building simulation model for DFIG using MPC-WM 173
11.3.2 Building simulation model for DFIG using MPRC-WM 178
11.4 Building simulation model for three-phase power CCG 184
11.4.1 Building simulation MPC for power converter 185
11.5 Summary 190

12. DFIG and three-phase power CCG experimental setup

Alfeu J. Sguarezi Filho

12.1 Experimental setups 191
12.1.1 DFIG setup 191

12.1.2 Three-phase CCG setup 191
12.1.3 The PLL setup 192
12.1.4 Data acquisition, power supply, and DC motor 192
12.1.5 System initialization 194
12.2 Information about the microcontroller 195
12.2.1 Functionality of the microcontroller 195
12.3 Predictive control implementation 196
12.3.1 Building embedded controller in DSP 196
12.4 Summary 198

A. DFIG parameters

B. Three-phase power CCG parameters

C. DC link voltage representation

Bibliography 205
Index 213

List of figures

Fig. 1.1	Basic structure diagram of a DFIG connected to the electrical grid.	2
Fig. 1.2	Structure block diagram of the book.	3
Fig. 2.1	The stationary and synchronous reference frames for the space vector.	7
Fig. 2.2	Induction machine representation in a three-phase system.	8
Fig. 2.3	Detailed circuit of CCG.	15
Fig. 2.4	Simplified circuit of CCG.	15
Fig. 2.5	Block diagram for implementation of the SPWM.	19
Fig. 2.6	SPWM curves: (a) comparison of the carrier and the modulating signals, (b) v_{a0}, (c) v_{b0}, and (d) v_{ab}.	19
Fig. 2.7	Voltage by using third harmonic component.	20
Fig. 2.8	THSPWM diagram.	20
Fig. 2.9	Space vector representation by using the switches states in the $\alpha\beta$ frame.	21
Fig. 2.10	SVM digital implementation.	24
Fig. 3.1	DFIG connected to the mains block diagram.	27
Fig. 3.2	Orientation diagrams: (a) voltage orientation and (b) stator flux orientation.	30
Fig. 3.3	DFIG rotor current control diagram.	32
Fig. 3.4	DFIG deadbeat rotor current control diagram.	35
Fig. 3.5	Detail of deadbeat diagram.	36
Fig. 3.6	DFIG deadbeat direct power control general scheme.	38
Fig. 3.7	Three-phase power CCG closed-loop grid current control by using PI controllers.	41
Fig. 4.1	MPC basic diagram.	44
Fig. 5.1	Graphical representation of converter switching conditions and the rotor voltage vector in the rotor reference.	59
Fig. 5.2	DPC for DFIG using modulated FCS-MPC block scheme of the implemented control algorithm.	60
Fig. 5.3	DPC for DFIG using modulated FCS-MPC block scheme of the rotor voltage prediction.	61
Fig. 5.4	Modulated FCS-MPC DPC. Step tests for apparent power: (a) active power and (b) reactive power.	63
Fig. 5.5	Modulated FCS-MPC DPC. Detail of the step test for apparent power: (a) active power and (b) reactive power.	64
Fig. 5.6	Modulated FCS-MPC DPC. Detail of the step test for apparent power: (a) reactive power and active power, (b) stator current and voltage.	64
Fig. 5.7	Modulated FCS-MPC DPC. Detail of the step test for apparent power: (a) reactive power and active power, (b) stator current and voltage.	64
Fig. 5.8	Modulated FCS-MPC DPC. THD of the rotor voltage during the test.	65
Fig. 5.9	Modulated FCS-MPC DPC. Rotor speed.	65

Fig. 5.10	Modulated FCS-MPC DPC. Step test for apparent power: (a) stator active power and (b) stator reactive power.	66
Fig. 5.11	Modulated FCS-MPC DPC. Rotor current during several-speed operation.	66
Fig. 6.1	Smart grid infrastructure.	68
Fig. 6.2	VSI sextant.	70
Fig. 6.3	Block diagram for the modulated FCS-MPC DPC for CCG.	74
Fig. 6.4	Block diagram for the wireless communication system.	74
Fig. 6.5	Experimental setup for tests.	76
Fig. 6.6	Response of received power step test, P_{gref} and Q_{gref}, using OFDM-CC.	77
Fig. 6.7	Grid voltage and current signals using OFDM-CC and response of received P_{gref} step test using OFDM-CC.	78
Fig. 6.8	Grid current signals using OFDM-CC ($Q = -300$ var). The blue (gray in print version) line is the reference signal and the dashed red (dark gray in print version) line is the current signal.	79
Fig. 6.9	Response of received power step test, Q_{gref} and P_{gref}, using OFDM-LDPC.	79
Fig. 6.10	Grid voltage and current signals using OFDM-LDPC and response of received P_{gref} step test using OFDM-LDPC.	80
Fig. 6.11	Grid current signals using OFDM-CC ($Q_{gref} = -300$ var).	81
Fig. 6.12	FFT analysis for switching.	81
Fig. 6.13	Current FFT analysis for OFDM-LDPC.	82
Fig. 6.14	Current FFT analysis for OFDM-CC.	82
Fig. 7.1	DFIG rotor current control by using MPC-WM diagram.	88
Fig. 7.2	MPC-WM. Step tests for rotor current: (a) q – quadrature axis, (b) d – direct axis, (c) in the $\alpha\beta$ frame, and (d) in the three-phase rotor frame.	90
Fig. 7.3	MPC-WM. Step tests for rotor current in the dq frame.	90
Fig. 7.4	MPC-WM. Detail of step test for rotor current in the dq frame.	91
Fig. 7.5	MPC-WM. Stator voltage and current during step of rotor current $i_{r,d}$: (a) stator voltage and current, (b) rotor current vector components.	91
Fig. 7.6	MPC-WM: Stator voltage and current during step of rotor current $i_{r,q}$. (a) Stator voltage and current. (b) Rotor current vector components.	92
Fig. 7.7	MPC-WM: Several speed operation test. (a) Speed of the rotor. (b) Rotor current in rotor frame. (c) Rotor current vector components in dq frame.	92
Fig. 7.8	MPC-WM: DFIG DPC.	96
Fig. 7.9	DPC MPC-WM. Step tests for active and reactive power: (a) active power, (b) reactive power.	97
Fig. 7.10	DPC MPC-WM. Detail of step test for active power: (a) active power, (b) reactive power.	97
Fig. 7.11	DPC MPC-WM. Stator voltage and current during step of active power: (a) reactive and active power, (b) stator voltage and current.	98
Fig. 7.12	DPC MPC-WM. Stator voltage and current during step of reactive power: (a) reactive and active power, (b) stator voltage and current.	98
Fig. 7.13	DPC MPC-WM. Several-speed operation test: (a) rotor speed, (b) rotor current, (c) reactive and active power.	99
Fig. 7.14	MPC-WM. CCG current control diagram.	102
Fig. 7.15	MPC-WM. Simulation result of the grid current behavior during the power step tests.	103
Fig. 7.16	MPC-WM: THD of the grid current.	103
Fig. 7.17	MPC-WM. Step tests for grid power control: (a) active power, (b) reactive power.	104
Fig. 7.18	MPC-WM. Detail of grid active power step: (a) active power, (b) reactive power.	104

Fig. 7.19	MPC-WM. Grid current during power step tests: (a) real component of the grid current vector, (b) imaginary component of the grid current vector.	105
Fig. 7.20	MPC-WM. Grid current during the reactive power step: (a) active and reactive power, (b) grid voltage and current.	105
Fig. 7.21	MPC-WM. Grid current during the active power step: (a) active and reactive power, (b) grid voltage and current.	106
Fig. 8.1	Control system loop by using the IMP.	108
Fig. 8.2	RC in a plug-in scheme.	109
Fig. 8.3	MPRC diagram.	114
Fig. 9.1	Diagram of rotor current control of DFIG using MPRC-WM.	127
Fig. 9.2	MPRC-WM. Step tests for rotor current vector components.	128
Fig. 9.3	MPRC-WM. Details of the step test for $i_{r,q}$.	129
Fig. 9.4	MPRC-WM. Detail of the step test for $i_{r,q}$: (a) stator voltage and current, (b) rotor current vector components.	129
Fig. 9.5	MPRC-WM. Details of the step test for $i_{r,d}$: (a) stator voltage and current, (b) rotor current vector components.	130
Fig. 9.6	MPRC-WM. Test using several-speed operation: (a) speed of the rotor, (b) rotor current in rotor frame, and (c) rotor current vector components in the dq frame.	130
Fig. 9.7	MPRC-WM. Test under voltage distorted by 7% of the 5th harmonic component: (a) stator voltage and current, (b) rotor current vector components.	131
Fig. 9.8	MPRC-WM. THD and harmonics components of stator current.	131
Fig. 9.9	MPC-WM. Test under voltage distorted by 7% of the 5th harmonic component: (a) stator voltage and current, (b) rotor current vector components.	132
Fig. 9.10	MPC-WM. THD and harmonics components of the stator current.	132
Fig. 9.11	Diagram of grid current control using MPRC-WM.	138
Fig. 9.12	MPRC-WM. Step tests for active and reactive power: (a) active power, (b) reactive power.	140
Fig. 9.13	MPRC-WM. Details of the step test for active power: (a) active power, (b) reactive power.	140
Fig. 9.14	MPRC-WM. Step tests for grid current vector components: (a) real component of grid current vector, (b) imaginary component of grid current vector.	141
Fig. 9.15	MPRC-WM. Grid voltage and current action during step test of active power: (a) reactive and active power, (b) grid voltage and current.	141
Fig. 9.16	MPRC-WM. Grid voltage and current action during a step test of reactive power: (a) reactive and active power, (b) grid voltage and current.	142
Fig. 9.17	MPRC-WM. Grid voltage and current action during a step test of active power during the distorted grid voltage: (a) reactive and active power, (b) grid voltage and current.	143
Fig. 9.18	MPRC-WM. THD of stator current during the distorted grid voltage.	143
Fig. 9.19	MPC-WM. Grid voltage and current action during a step test of active power during the distorted grid voltage: (a) reactive and active power, (b) grid voltage and current.	144
Fig. 9.20	MPC-WM. THD of stator current during the distorted grid voltage.	144
Fig. 10.1	Basic structure of the PLL block diagram.	145
Fig. 10.2	PLL for three-phase systems' block diagram.	147
Fig. 10.3	Experimental result of SRF-PLL test for unbalanced voltage sag: (a) three-phase voltage, (b) grid voltages elements in the dq frame, (c) grid frequency, and (d) grid angle.	149
Fig. 10.4	Block diagram of structure of the DSOGI.	150

Fig. 10.5 Diagram of the DSOGI-MAF-FPS-PLL. 154

Fig. 10.6 Experimental result of DSOGI-MAF-FPS-PLL test for unbalanced sag: (a) three-phase voltage, (b) positive sequence of the grid voltages elements in the dq frame, (c) grid frequency, and (d) grid angle. 155

Fig. 10.7 Experimental result of DSOGI-MAF-FPS-PLL test for voltage sag with the 5th harmonic: (a) three-phase voltage, (b) positive sequence of the grid voltages elements in the dq frame, (c) grid frequency, and (d) grid angle. 156

Fig. 10.8 Experimental result of DSOGI-MAF-FPS-PLL test under distorted voltage: (a) three-phase voltage, (b) positive sequence of the grid voltages elements in the dq frame, (c) grid frequency, and (d) grid angle. 156

Fig. 11.1 Clarke transformation (CT). 158

Fig. 11.2 Park transformation (PT). 158

Fig. 11.3 Park transformation (PT) in Simulink. 159

Fig. 11.4 Inverse Clarke transformation. 159

Fig. 11.5 Inverse Park transformation. 160

Fig. 11.6 PWM in Simulink. 161

Fig. 11.7 SVM in Simulink. 161

Fig. 11.8 Model for three-phase power converter in Simulink. 168

Fig. 11.9 Three-phase grid voltages model. 168

Fig. 11.10 Currents' block of DFIG simulation model. 170

Fig. 11.11 Fluxes' block of DFIG simulation model. 171

Fig. 11.12 Torque block of DFIG simulation model. 172

Fig. 11.13 DFIG system simulation block. 172

Fig. 11.14 Model predictive control block, MATLAB/Simulink implementation. 174

Fig. 11.15 Predictive Repetitive control block, MATLAB/Simulink implementation. 179

Fig. 11.16 Complete three-phase power CCG model in Simulink. 185

Fig. 11.17 Current reference in Simulink. 187

Fig. 11.18 Model predictive control block, MATLAB/Simulink implementation. 187

Fig. 12.1 The experimental setup of DFIG used in MPC tests. 191

Fig. 12.2 The experimental setup of three-phase CCG used in MPC tests. 192

Fig. 12.3 The experimental setup for PLL tests. 193

List of tables

Table 2.1 Switching states and voltage vectors. 22
Table 2.2 PWM time scale definition. 23
Table 2.3 Times for registers in function of the sector. 24
Table 3.1 Doubly-fed induction machine operation conditions. 28
Table 5.1 Switching states and voltage vectors. 58
Table 5.2 Sector of reference rotor voltage vector. 58
Table 5.3 Calculation of the duty cycle of the adjacent vectors to each sector $k - mfcs$ in the $\alpha\beta r$ frame. 63
Table 6.1 Switching states and voltage vectors. 69
Table 9.1 Distorted voltage harmonics components. 142
Table A.1 IM parameters. 199
Table B.1 Three-phase power CCG parameters. 201

List of contributors

Carlos E. Capovilla, Federal University of ABC, Santo André, SP, Brazil

Ivan R.S. Casella, Federal University of ABC, Santo André, SP, Brazil

Eliomar R. Conde D., Federal University of ABC, Santo André, SP, Brazil

Rogério V. Jacomini, Federal Institute of São Paulo, Hortolândia, SP, Brazil

Fernando Lino, Federal University of ABC, Santo André, SP, Brazil

Angelo S. Lunardi, Federal University of ABC, Santo André, SP, Brazil

Alfeu J. Sguarezi Filho, Federal University of ABC, Santo André, SP, Brazil

Biography

Professor Dr. Alfeu J. Sguarezi Filho

He was born in Cuiabá – MT – Brazil in 1981, received his Electrical Eng. degree from ÁREA1 – Faculty of Engineering in Ba – Brazil in 2005, his Master's and PhD degrees from Campinas University in SP – Brazil in 2007 and 2010, respectively. He is IEEE Senior member. Since 2012, he is a full time professor at Federal University of ABC – UFABC, in Santo André – SP – Brazil, teaching in the areas of Electrical Machines, Power Electronics, and Electrical Drives.

Abbreviation list

Abbreviations

AC	Alternating Current
CCG	Converter Connected to the Grid
DC	Direct Current
DFIG	Doubly-Fed Induction Generator
DFT	Discrete Fourier Transform
DPC	Direct Power Control
DTC	Direct Torque Control
DSP	Digital Signal Processor
FCS	Finite Control Set
FOC	Field Oriented Control
IM	Induction Machine
IMP	Internal Model Principle
IGBT	Insulated-Gate Bipolar Transistor
MPC	Model Predictive Control
MPRC	Model Predictive Repetitive Control
PWM	Pulse Width Modulation
PI	Proportional-Integral
PLL	Phase Locked Loop
SPWM	Sinusoidal PWM
SVOC	Stator Voltage Oriented Control
SVM	Space Vector Modulation
THSPWM	Third Harmonic Sinusoidal PWM
RC	Repetitive Control
VSI	Voltage Source Inverter
WM	With Modulator
ZOH	Zero Order Hold

Variables and symbols

$\mathcal{I}m$	Imaginary component of the complex number
$\mathcal{R}e$	Real component of the complex number
i	Instantaneous value of current as a function of time
l	Related to the harmonic components
s	Laplace operator
v	Instantaneous value of voltage as a function of time
$\vec{v}$	Voltage vector
$\vec{i}$	Current vector
$\vec{\lambda}$	Flux vector
J	Cost function

T_L	Load or mechanical torque
ω	Speed or angular frequency
NP	Pole pairs
n_y	Prediction horizon
n_u	Control horizon
R	Resistance
L	Inductance
L_m	Mutual inductance
σ	Global leakage
T_e	Electromagnet torque
JJ	Total moment of inertia
x	State
Δu	Increment of input
λ_s, $\|\lambda_s\|$	Magnitude of stator flux
rms	Root mean square value
q^{-1}	Backward shift operator
k	Instant sampling
$(k+1)$	Next sampling
I	Identity matrix
O	Zero matrix
θ	Angular position
P	Active power
Q	Reactive power
W_y	Weighting matrix for prediction effort
W_u	Weighting matrix for control effort
u	Input
out	Number of outputs

Superscript

"*"	Conjugate of the complex number

Subscripts

s	Related to the stator
r	Related to the rotor
g	Related to the electrical grid
$\alpha\beta$	Related to the stationary reference frame
α	Related to the real component of the stationary reference frame
β	Related to the imaginary component of the stationary reference frame
$\alpha\beta r$	Related to the rotor reference frame
αr	Related to the real component of the rotor reference frame
βr	Related to the imaginaty component of the rotor reference frame
ρ	Related to the filter order
dq	Related to the synchronous reference frame
d	Related to the real component of the synchronous reference frame
q	Related to the imaginaty component of the synchronous reference frame
ee	Related to the expanded model
ir	Related to the rotor current
mec	Related to the mechanical components of the machine
pq	Related to the stator active and reactive power
ref	Related to the references
sl	Related to the slip

Chapter 1

Introduction

Alfeu J. Sguarezi Filho
Federal University of ABC, Santo André, SP, Brazil

1.1 Overview

The evolution of power semiconductor components permitted processing power at levels of thousands of volts and ampers. In this context, several power electronics devices built with arrangements of these semiconductor elements enabled the conversion of energy from AC–AC, with different frequencies and amplitudes, DC–AC, AC–DC, and DC–AC [1]. Thus, it was possible to make the different ways of using electric energy more flexible and, consequently, to diversify the ways of its application. Finally, the evolution of power semiconductors, the increase in the processing and storage capacity of processors and their peripherals, respectively, enabled greater flexibility in the implementation of more complex control algorithms.

One application of power electronics is in renewable energy such as wind, small hydroelectric plants, or in electrical energy storage systems, among others [2]. This type of application is important due to the growing demand for energy in developed and developing countries, and in the search for alternative energy sources rather than using fossil fuels. In this case, power electronics can be used to process energy from electric generators or to process energy present in the energy storage systems.

Among the generators employed in the renewable energy, the doubly-fed induction generator (DFIG) has become popular. This type of generator emerged as an alternative to permanent magnets synchronous generators, due to its low cost and robustness. The ordinary configuration of DFIG employed in renewable energy systems has its stator connected directly to the grid and its rotor connected to the grid through a bidirectional converter, which processes a maximum of 30% of the total power of the generator [3]. In this way, it decreases the cost of the system. The bidirectional converter, called back-to-back converter, is an arrangement of two electronic power converters that have AC input/output and share the same DC link. A diagram of the DFIG using this configuration is depicted in Fig. 1.1. It is possible to notice the basic structure of the grid and rotor side converters control using the measurements of current, torque, power or speed of the generator, or using the measurements of current, voltage or power of the grid.

Model Predictive Control for Doubly-Fed Induction Generators and Three-Phase Power Converters
https://doi.org/10.1016/B978-0-32-390964-8.00010-5

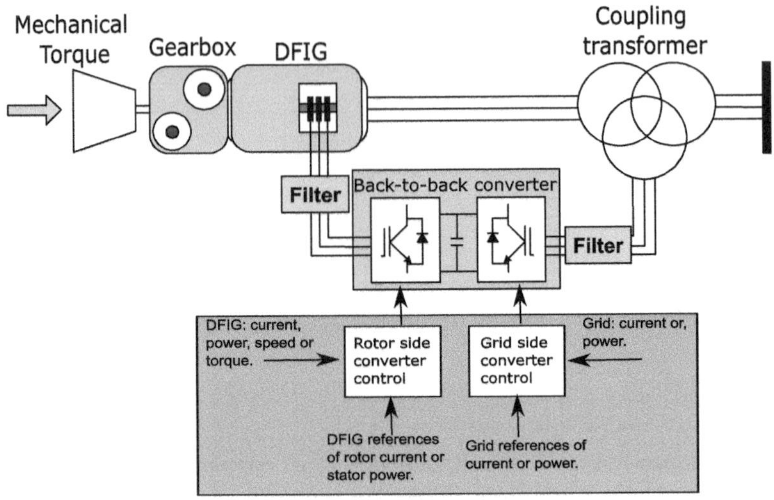

FIGURE 1.1 Basic structure diagram of a DFIG connected to the electrical grid.

From the control techniques' point of view, these can be performed in closed loop, with the use of hysteresis, linear, nonlinear or intelligent controllers, together with the field-oriented technique applied in the control of speed, torque, currents or power, in order to improve the performance of the system. Another control technique that has been widely used in power electronics is predictive control.

Predictive control was developed between the 1960s and 1970s and its application in power electronics started in the 1980s. This type of control is a set of controllers that basically predict the future behavior of the plant and, starting from this information and the minimized cost function, calculates the input to be applied in the system so that the references are reached [4,5]. Thus, predictive control techniques can represent plants by transfer functions, impulse response functions, space state equation or artificial intelligence, among others. From the future behavior of the plant and using the cost function, the input is calculated. And, this can be the states of each switch of the power electronic converter in which modulation is not used, or it can be a voltage or current vector operating in conjunction with a modulator to set the converter switches.

This book introduces the design of predictive controllers for the converter connected to the rotor of the DFIG and the power converter connected to the electrical grid. The controllers are designed with their dynamic mathematical models using the representation by the discrete state space equations. By predicting the behavior of the generator and converter linked to the network, one calculates the state of the switches of the converter, or the voltage vector is selected together with a modulation technique using the minimized cost function.

1.2 Structure of the book

After the introduction presented in this chapter, Chapter 2 introduces the representation of the mathematical transformations of coordinate systems and the dynamic equations of the induction machine and the three-phase converter connected to the electrical grid. In addition, sinusoidal and spatial vector modulation techniques are presented.

The fundamental concepts of vector control for DFIG and the converter connected to the power grid are presented in Chapter 3. In addition, closed-loop current control operation using proportional-integral and deadbeat controllers is introduced.

Chapter 4 introduces the fundamental concepts of the model predictive control finite control set type and model predictive control with the use of a modulator. The main information covers the prediction models, cost functions, and constraints of the predictive control. The design of the controller with the use of a modulator for the rotor current control of the DFIG and the current control of the converter connected to the grid are presented in Chapter 7. The direct power control for DFIG using the modulated model predictive control is introduced in Chapter 5. Finally, an application in smart grid environment using wireless coded modulated finite control set direct power control for renewable energy sources are described in Chapter 6. Simulation and experimental results corroborate the performance of the model predictive controllers.

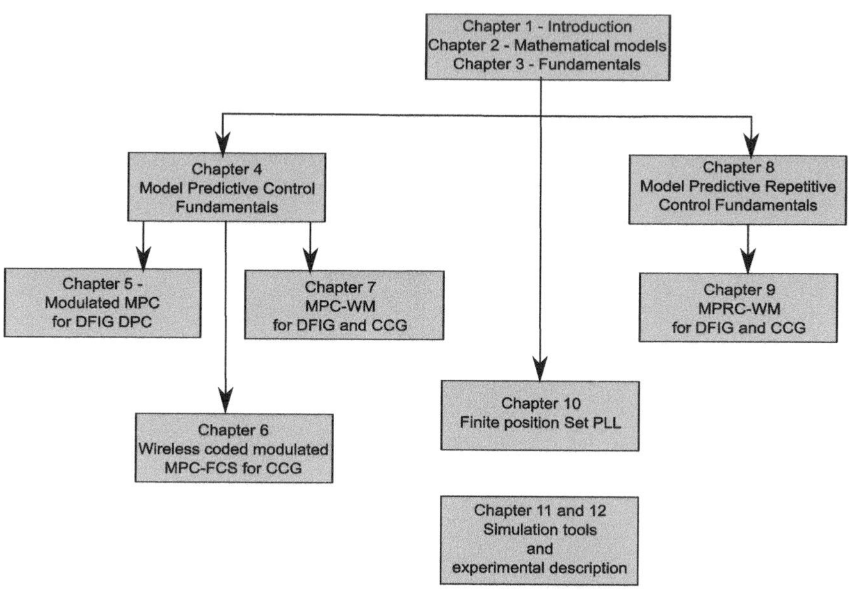

FIGURE 1.2 Structure block diagram of the book.

The fundamental concepts of the repetitive predictive controller are presented in Chapter 8. The rotor current of the DFIG control and the current control of the converter connected to the network design applying the mentioned model predictive repetitive controller with the use of a modulator to control are shown in Chapter 9. Additional pieces of information about the fundamentals of repetitive control are presented. Simulation and experimental results corroborate the performance of the model predictive repetitive controllers including the operation when the grid voltages are distorted by harmonics components.

Chapter 10 introduces the concepts of phase-locked loop applied to synchronization with the electrical grid and the use of model predictive control concepts in this type of application when the electrical grid operates under distorted voltage. The results corroborate the performance of phase-locked loop.

Finally, the description of the simulations performed and the experimental benches built in the Laboratory of Power Electronics and Smart Grids of UFABC are presented in Chapters 11 and 12, respectively.

The structure of the book block diagram in which one sees the links between the chapters is depicted in Fig. 1.2.

Chapter 2

Induction machine and three-phase power converter dynamic models

Alfeu J. Sguarezi Filho
Federal University of ABC, Santo André, SP, Brazil

2.1 Space vector notation

The concept of a space vector [6] is employed in the mathematical representation of the state variables of the induction machines (IM), such as voltage, current, and fluxes. In this way, the equations that represent the machine become more compact and easier to interpret. This concept allows the representation of the three-phase system by means of an orthogonal system. The space vector is defined as

$$\vec{x} = \frac{2}{3}\left[x_a(t) + a x_b(t) + a^2 x_c(t)\right], \tag{2.1}$$

where x_a, x_b, and x_c represent the IM phase quantities mentioned, and they satisfy the condition $x_a + x_b + x_c = 0$. In this way, $a = e^{j120^\circ}$, $j = \sqrt{-1}$ is the complex operator, or by using Euler's identity $a = \cos(120^\circ) + j\sin(120^\circ)$, $a^2 = e^{j240^\circ} = \cos(240^\circ) + j\sin(240^\circ)$. The constant $\frac{2}{3}$ is a normalization factor, which means that the values are in accordance with the peak values of each variable. Hence, this assumption is interesting in power electronics field due to the fact the controllers can compute peak values and, also, they are interesting to the PWM algorithms.

2.1.1 The stationary reference frame ($\alpha\beta$)

The projection of the space vector $\vec{x}$ into the real α and imaginary β axes is fixed. Thus, $\vec{x} = x_\alpha + jx_\beta$. The space vector representation for the voltage vector $\vec{v}_{\alpha\beta} = v_\alpha + jv_\beta$ in the stationary reference frame by using three-phase voltages v_{abc} can be expressed by using the Clarke's transformation [7]:

$$\vec{v}_{\alpha\beta} = [T_{abc}]v_{abc} \tag{2.2}$$

Model Predictive Control for Doubly-Fed Induction Generators and Three-Phase Power Converters
https://doi.org/10.1016/B978-0-32-390964-8.00011-7

and

$$\begin{bmatrix} v_\alpha \\ v_\beta \end{bmatrix} = \begin{bmatrix} 1 & -\frac{1}{2} & -\frac{1}{2} \\ 0 & \frac{\sqrt{3}}{2} & -\frac{\sqrt{3}}{2} \end{bmatrix} \begin{bmatrix} v_a \\ v_b \\ v_c \end{bmatrix}. \tag{2.3}$$

Conversely, it is possible to calculate the three-phase voltages by using the voltage vector $\vec{v}_{\alpha\beta}$ which can be seen in:

$$v_{abc} = [T_{abc}]^{-1}\vec{v}_{\alpha\beta}, \tag{2.4}$$

$$\begin{bmatrix} v_a \\ v_b \\ v_c \end{bmatrix} = \begin{bmatrix} 1 & 0 \\ -\frac{1}{2} & \frac{\sqrt{3}}{2} \\ -\frac{1}{2} & -\frac{\sqrt{3}}{2} \end{bmatrix} \begin{bmatrix} v_\alpha \\ v_\beta \end{bmatrix}. \tag{2.5}$$

The mentioned transformations can be employed to any current, fluxes, or voltages.

2.1.2 The synchronous reference frame (dq)

The synchronous reference frame (dq) rotates at frequency ω, and it is located at the $\theta = \omega \cdot t$ position on the stationary reference frame ($\alpha\beta$). In this way, d is the real axis and q is the imaginary axis of the dq reference frame. The representation for the voltage vector $\vec{v}_{dq} = v_d + jv_q$ by using voltage vector $\vec{v}_{\alpha\beta}$ can be expressed by applying the Park's transformation [8]:

$$\vec{v}_{dq} = [T_{dq}]\vec{v}_{\alpha\beta} \tag{2.6}$$

and

$$\begin{bmatrix} v_d \\ v_q \end{bmatrix} = \begin{bmatrix} \cos(\theta) & \sin(\theta) \\ -\sin(\theta) & \cos(\theta) \end{bmatrix} \begin{bmatrix} v_\alpha \\ v_\beta \end{bmatrix}. \tag{2.7}$$

In the same way, Eq. (2.6) can be represented as

$$\vec{v}_{dq} = v_d + jv_q = \vec{v}_{\alpha\beta}\, e^{-j\theta} = \left[v_\alpha + jv_\beta\right][\cos(\theta) - j\sin(\theta)], \tag{2.8}$$

which means for each axis:

$$\begin{aligned} v_d &= v_\alpha \cos(\theta) + v_\beta \sin(\theta), \\ v_q &= -v_\alpha \sin(\theta) + v_\beta \cos(\theta). \end{aligned} \tag{2.9}$$

The result presented in Eq. (2.9) is the same as when Eq. (2.7) is employed.

Conversely, it is possible to calculate the voltage vector $\vec{v}_{\alpha\beta}$ by using the vector $\vec{v}_{dq}$ and θ as can be seen in:

$$\vec{v}_{\alpha\beta} = [T_{dq}]^{-1}\vec{v}_{dq} \tag{2.10}$$

and

$$\begin{bmatrix} v_\alpha \\ v_\beta \end{bmatrix} = \begin{bmatrix} \cos(\theta) & -\sin(\theta) \\ \sin(\theta) & \cos(\theta) \end{bmatrix} \begin{bmatrix} v_d \\ v_q \end{bmatrix}. \tag{2.11}$$

In the same way, Eq. (2.10) can be represented as

$$\vec{v}_{\alpha\beta} = v_\alpha + j v_\beta = \vec{v}_{dq}\, e^{j\theta} = \left[v_d + j v_q\right][\cos(\theta) + j\sin(\theta)], \tag{2.12}$$

which means for each axis:

$$\begin{aligned} v_\alpha &= v_d \cos(\theta) - v_q \sin(\theta), \\ v_\beta &= v_d \sin(\theta) + v_q \cos(\theta). \end{aligned} \tag{2.13}$$

The result presented in Eq. (2.13) is the same as when Eq. (2.11) is employed.

Again, the mentioned transformations can be employed to any current, fluxes, or voltages.

Fig. 2.1 depicts the stationary and synchronous reference frames for the space vector $\vec{v}$.

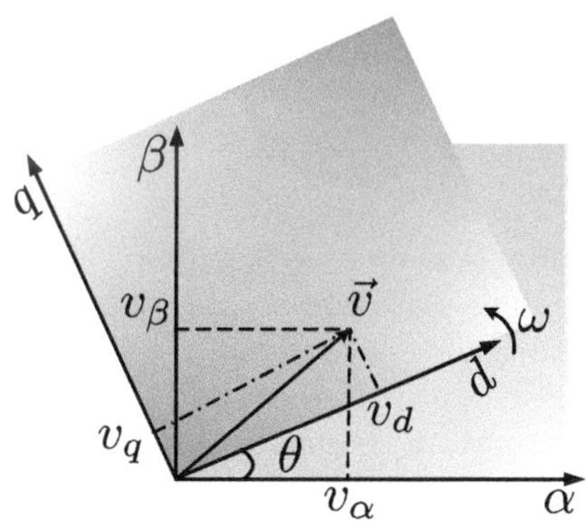

FIGURE 2.1 The stationary and synchronous reference frames for the space vector.

2.2 Induction machine dynamic model

The induction machine has three-phase windings, characterized by coils, spatially displaced by 120° electric degrees. The magnetic flux of the air gap has a radial direction. In this case, the surfaces of the stator and rotor are smooth, the permeability of iron is considered infinite, and, disregarding the effects on the extremities, the magnetic field becomes dimensional.

A typical diagram of the induction machine is shown in Fig. 2.2.

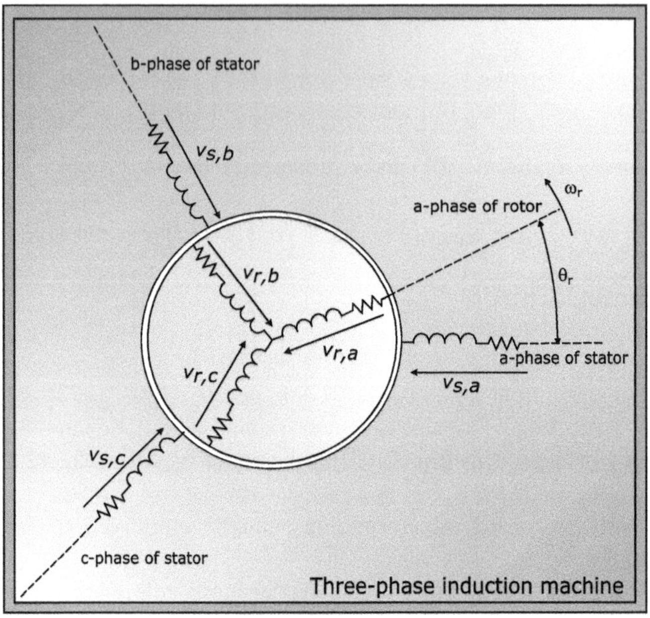

FIGURE 2.2 Induction machine representation in a three-phase system.

2.2.1 IM representation in three-phase systems

For the representation of the dynamic behavior of the IM, some considerations are assumed, namely [9]:

- The permeability of iron is assumed to be infinite;
- It is assumed that the three-phase system is symmetrical;
- The machine is considered magnetically linear;
- Skin effect and losses in iron are disregarded;
- Resistances and reactances are constant;
- In a steady state, the voltages and currents are sinusoidal.

Thus, with these considerations in mind and, based on the model presented in Fig. 2.2, the stator and rotor expressions per phase are described, respectively, as:

$$v_{s,a} = i_{s,a} R_s + \frac{d\lambda_{s,a}}{dt}, \tag{2.14}$$

$$v_{s,b} = i_{s,b} R_s + \frac{d\lambda_{s,b}}{dt}, \tag{2.15}$$

$$v_{s,c} = i_{s,c} R_s + \frac{d\lambda_{s,c}}{dt}, \tag{2.16}$$

$$v_{r,a} = i_{r,a} R_r + \frac{d\lambda_{r,a}}{dt}, \tag{2.17}$$

$$v_{r,b} = i_{r,b} R_r + \frac{d\lambda_{r,b}}{dt}, \tag{2.18}$$

$$v_{r,c} = i_{r,c} R_r + \frac{d\lambda_{r,c}}{dt}, \tag{2.19}$$

where v represents the voltage, i expresses the current, R_s is the stator resistance, R_r is the rotor resistance, λ is the flux, and the subscripts a, b, and c represent each phase of the three-phase system, and the subscripts s and r represent the stator and rotor, respectively.

The relationship between the rotor angular frequency and the stator angular frequency (ω_s) is given by

$$\omega_s = \omega_r + \omega_{sl}, \tag{2.20}$$

where ω_{sl} is the slip angular frequency and ω_r is the angular frequency of the machine. The mechanical speed ω_{mec} can be calculated by means of ω_r and the pole pairs NP of the machine:

$$\omega_r = NP\omega_{mec}. \tag{2.21}$$

2.2.2 IM representation in stationary reference frame ($\alpha\beta$)

The equations for the stator and rotor voltages of the induction machine are described by means of the space vector presented in Sections 2.1 and 2.1.1 by using Eqs. (2.14)–(2.19). Hence, they can be represented as follows:

The stator equations are

$$\vec{v}_{s,\alpha\beta} = \vec{i}_{s,\alpha\beta} R_s + \frac{d\vec{\lambda}_{s,\alpha\beta}}{dt}, \tag{2.22}$$

which means

$$v_{s,\alpha} = i_{s,\alpha} R_s + \frac{d\lambda_{s,\alpha}}{dt} \tag{2.23}$$

and

$$v_{s,\beta} = i_{s,\beta} R_s + \frac{d\lambda_{s,\beta}}{dt}; \tag{2.24}$$

The rotor equations are

$$\vec{v}_{r,\alpha\beta r} = \vec{i}_{r,\alpha\beta r} R_r + \frac{d\vec{\lambda}_{r,\alpha\beta r}}{dt}, \tag{2.25}$$

which means

$$v_{r,\alpha r} = i_{r,\alpha r} R_r + \frac{d\lambda_{r,\alpha r}}{dt} \tag{2.26}$$

and

$$v_{r,\beta r} = i_{r,\beta r} R_r + \frac{d\lambda_{r,\beta r}}{dt}. \tag{2.27}$$

Note that Eq. (2.22) is fixed in the stationary reference frame ($\alpha\beta$) and Eq. (2.25) is fixed in the rotor reference frame ($\alpha\beta r$).

The relationship between the fluxes and current of stator and rotor is given by

$$\vec{\lambda}_s = L_s \vec{i}_{s,\alpha\beta} + L_m \vec{i}_{r,\alpha\beta r} e^{j\theta_r} = L_s \vec{i}_{s,\alpha\beta} + L_m \vec{i}_{r,\alpha\beta} \tag{2.28}$$

and

$$\vec{\lambda}_{r,\alpha\beta r} = L_r \vec{i}_{r,\alpha\beta r} + L_m \vec{i}_{s,\alpha\beta} e^{-j\theta_r} = L_r \vec{i}_{r,\alpha\beta r} + L_m \vec{i}_{s,\alpha\beta r}. \tag{2.29}$$

It can be noted that Eqs. (2.28) and (2.29) employ the coordinating transformation presented in Eqs. (2.12) and (2.8), respectively. And, $\theta_r = \int \omega_r dt$.

In the above, $\vec{v}$, $\vec{i}$, $\vec{\lambda}$ are the voltage, current, and flux space vectors, respectively, R is the proper resistance, L the inductance, L_m the magnetization inductance, the subscripts s and r represent the stator and rotor, respectively. The proper inductance of stator L_s and rotor L_r can be expressed by means of the stator leakage inductance L_{ls} and the rotor leakage inductance L_{lr} and they can be calculated by the expressions:

$$\begin{aligned} L_s &= L_{ls} + L_m, \\ L_r &= L_{lr} + L_m. \end{aligned} \tag{2.30}$$

In the IM dynamic model in the stationary reference frame, the stator voltage vector represented by Eq. (2.22) remains the same. The rotor voltage vector is achieved by using the coordinating transformation (multiplying Eq. (2.25) by $e^{j\theta_r}$). Hence, the IM model in the stator reference $\alpha\beta$ frame can be modeled as

$$\vec{v}_{s,\alpha\beta} = \vec{i}_{s,\alpha\beta} R_s + \frac{d\vec{\lambda}_{s,\alpha\beta}}{dt}, \tag{2.31}$$

which means:

$$\begin{aligned} v_{s,\alpha} &= i_{s,\alpha} R_s + \frac{d\lambda_{s,\alpha}}{dt} \\ v_{s,\beta} &= i_{s,\beta} R_s + \frac{d\lambda_{s,\beta}}{dt} \end{aligned} \tag{2.32}$$

and the rotor equations can be represented as follows:

$$\vec{v}_{r,\alpha\beta} = \vec{i}_{r,\alpha\beta} R_r + \frac{d\vec{\lambda}_{r,\alpha\beta}}{dt} - j\omega_r \vec{\lambda}_{r,\alpha\beta} \tag{2.33}$$

which means:

$$v_{r,\alpha} = i_{r,\alpha} R_r + \frac{d\lambda_{r,\alpha}}{dt} + \omega_r \lambda_{r,\beta},$$
$$v_{r,\beta} = i_{r,\beta} R_r + \frac{d\lambda_{r,\beta}}{dt} - \omega_r \lambda_{r,\alpha}. \tag{2.34}$$

The representation of the fluxes vectors can be performed as per the current vector as follows:

$$\vec{\lambda}_{s,\alpha\beta} = L_s \vec{i}_{s,\alpha\beta} + L_m \vec{i}_{r,\alpha\beta},$$
$$\lambda_{s,\beta} = L_s i_{s,\alpha} + L_m i_{r,\alpha},$$
$$\lambda_{s,\alpha} = L_s i_{s,\beta} + L_m i_{r\beta}, \tag{2.35}$$

and

$$\vec{\lambda}_{r,\alpha\beta} = L_r \vec{i}_{r,\alpha\beta} + L_m \vec{i}_{s,\alpha\beta},$$
$$\lambda_{r,\alpha} = L_r i_{r,\alpha} + L_m i_{s,\alpha},$$
$$\lambda_{r,\beta} = L_r i_{r,\beta} + L_m i_{s,\beta}, \tag{2.36}$$

where $\vec{v}_{s,\alpha\beta} = v_{s,\alpha} + j v_{s,\beta}$, $\vec{i}_{s,\alpha\beta} = i_{s,\alpha} + j i_{s,\beta}$, $\vec{\lambda}_{s,\alpha\beta} = \lambda_{s,\alpha} + j\lambda_{s,\beta}$, $\vec{v}_{r,\alpha\beta} = v_{r,\alpha} + j v_{r,\beta}$, $\vec{i}_{r,\alpha\beta} = i_{r,\alpha} + j i_{r,\beta}$, and $\vec{\lambda}_{r,\alpha\beta} = \lambda_{r,\alpha} + j\lambda_{r,\beta}$.

In Eq. (2.33), it is necessary to note the expression below:

$$\frac{d\vec{\lambda}_{r,\alpha\beta}}{dt} = \frac{d(\vec{\lambda}_{r,\alpha\beta r} \cdot e^{j\theta_r})}{dt} = \frac{d\vec{\lambda}_{r,\alpha\beta r}}{dt} \cdot e^{j\theta_r} + j \underbrace{\frac{d\theta_r}{dt}}_{\omega_r} \underbrace{\vec{\lambda}_{r,\alpha\beta r} \cdot e^{j\theta_r}}_{\vec{\lambda}_{r,\alpha\beta}}$$
$$\frac{d\vec{\lambda}_{r,\alpha\beta r}}{dt} e^{j\theta_r} = \frac{d\vec{\lambda}_{r,\alpha\beta}}{dt} - j\omega_r \vec{\lambda}_{r,\alpha\beta} \tag{2.37}$$

2.2.2.1 Electromagnetic torque and power equations

The electromagnetic torque can be represented by the equations

$$T_e = \frac{3}{2} NP\, \mathcal{I}m(\vec{i}_{s,\alpha\beta} \cdot \vec{\lambda}^*_{s,\alpha\beta}) = \frac{3}{2} NP(-i_{s,\alpha}\lambda_{s,\beta} + i_{s,\beta}\lambda_{s,\alpha}). \tag{2.38}$$

There are several ways to represent the T_e as

$$T_e = \frac{3}{2} NP \frac{L_m}{L_s} \mathcal{I}m(\vec{\lambda}_{s,\alpha\beta} \cdot \vec{i}^*_{r,\alpha\beta}) = \frac{3}{2} NP\, \mathcal{I}m(\vec{\lambda}_{r,\alpha\beta} \cdot \vec{i}^*_{r,\alpha\beta})$$
$$= \frac{3}{2} NP\, L_m\, \mathcal{I}m(\vec{i}_{s,\alpha\beta} \cdot \vec{i}^*_{r,\alpha\beta}), \tag{2.39}$$

where "*" represents the conjugate of the complex number and $\mathcal{I}m$ is the imaginary component of the complex number.

The apparent power $S_{s,\alpha\beta}$ can be represented as

$$S_{s,\alpha\beta} = \frac{3}{2}\left(\vec{v}_{s,\alpha\beta}\,\vec{i}^{\,*}_{s,\alpha\beta}\right). \tag{2.40}$$

The stator active $P_{s,\alpha\beta}$ and reactive $Q_{s,\alpha\beta}$ power are components of the apparent power:

$$P_{s,\alpha\beta} = P_s = \mathcal{R}e(S_{s,\alpha\beta}) = \frac{3}{2}\left(v_{s,\alpha}i_{s,\alpha} + v_{s,\beta}i_{s,\beta}\right), \tag{2.41}$$

$$Q_{s,\alpha\beta} = Q_s = \mathcal{I}m(S_{s,\alpha\beta}) = \frac{3}{2}\left(v_{s,\beta}i_{s,\alpha} - v_{s,\alpha}i_{s,\beta}\right), \tag{2.42}$$

where $\mathcal{I}m$ is the imaginary component and $\mathcal{R}e$ is the real component of the complex number.

In the same way, the apparent power $S_{r,\alpha\beta}$ of the rotor can be represented as

$$S_{r,\alpha\beta} = \frac{3}{2}\left(\vec{v}_{r,\alpha\beta}\,\vec{i}^{\,*}_{r,\alpha\beta}\right). \tag{2.43}$$

As mentioned, the rotor active $P_{r,\alpha\beta}$ and reactive $Q_{r,\alpha\beta}$ power are components of the apparent represented as

$$P_{r,\alpha\beta} = \mathcal{R}e(S_{r,\alpha\beta}) = \frac{3}{2}\left(v_{r,\alpha}i_{r,\alpha} + v_{r,\beta}i_{r,\beta}\right), \tag{2.44}$$

$$Q_{r,\alpha\beta} = \mathcal{I}m(S_{r,\alpha\beta}) = \frac{3}{2}\left(v_{r,\beta}i_{r,\alpha} - v_{r,\alpha}i_{r,\beta}\right). \tag{2.45}$$

2.2.3 IM representation in synchronous reference frame (dq)

IM dynamic model in the synchronous reference frame refers to the stator voltage vector represented by Eq. (2.31), and the rotor voltage vector (Eq. (2.33)) is achieved by using the coordinating transformation (multiplying Eqs. (2.31) and (2.33) by $e^{-j\theta_s}$). Hence, the IM model in the synchronous reference frame can be written as

$$\vec{v}_{s,dq} = \vec{i}_{s,dq}R_s + \frac{d\vec{\lambda}_{s,dq}}{dt} + j\omega_s\vec{\lambda}_{s,dq}, \tag{2.46}$$

which means

$$\begin{aligned} v_{s,d} &= i_{s,d}R_s + \frac{d\lambda_{s,d}}{dt} - \omega_s\lambda_{s,q}, \\ v_{s,d} &= i_{s,q}R_s + \frac{d\lambda_{s,q}}{dt} + \omega_s\lambda_{s,d}. \end{aligned} \tag{2.47}$$

The rotor model is given by

$$\vec{v}_{r,dq} = \vec{i}_{r,dq} R_r + \frac{d\vec{\lambda}_{r,dq}}{dt} + j(\omega_s - \omega_r)\vec{\lambda}_{r,dq}, \tag{2.48}$$

which means

$$\begin{aligned} v_{r,d} &= i_{r,d} R_r + \frac{d\lambda_{r,d}}{dt} - (\omega_s - \omega_r)\lambda_{r,q}, \\ v_{r,q} &= i_{r,q} R_r + \frac{d\lambda_{r,q}}{dt} + (\omega_s - \omega_r)\lambda_{r,d}. \end{aligned} \tag{2.49}$$

The fluxes vector using the rotor and stator current vectors can be obtained from

$$\begin{aligned} \vec{\lambda}_{s,dq} &= L_s \vec{i}_{s,dq} + L_m \vec{i}_{r,dq}, \\ \lambda_{s,d} &= L_s i_{s,d} + L_m i_{r,d}, \\ \lambda_{s,q} &= L_s i_{s,q} + L_m i_{r,q}, \end{aligned} \tag{2.50}$$

and

$$\begin{aligned} \vec{\lambda}_{r,dq} &= L_r \vec{i}_{r,dq} + L_m \vec{i}_{s,dq}, \\ \lambda_{r,d} &= L_r i_{r,d} + L_m i_{s,d}, \\ \lambda_{r,q} &= L_r i_{r,q} + L_m i_{s,q}, \end{aligned} \tag{2.51}$$

where $\vec{v}_{s,dq} = v_{s,d} + jv_{s,q}$, $\vec{i}_{s,dq} = i_{s,d} + ji_{s,q}$, $\vec{\lambda}_{s,dq} = \lambda_{s,d} + j\lambda_{s,q}$, $\vec{v}_{r,dq} = v_{r,d} + jv_{r,q}$, $\vec{i}_{r,dq} = i_{r,d} + ji_{r,q}$, and $\vec{\lambda}_{r,dq} = \lambda_{r,d} + j\lambda_{r,q}$.

In Eq. (2.48), it is necessary to note the expression below:

$$\frac{d\vec{\lambda}_{r,\alpha\beta}}{dt} e^{-j\theta_s} = \frac{d\vec{\lambda}_{r,\alpha\beta r}}{dt} e^{j\theta_r} e^{-j\theta_s} = \frac{d\vec{\lambda}_{r,dq}}{dt} + j(\omega_s - \omega_r)\vec{\lambda}_{r,dq} \tag{2.52}$$

2.2.3.1 Electromagnetic torque and power equations

The electromagnetic torque can be represented by the equations

$$T_e = \frac{3}{2} NP\, \mathcal{I}m(\vec{i}_{s,dq} \cdot \vec{\lambda}^*_{s,dq}) = \frac{3}{2} NP(-i_{s,d}\lambda_{s,q} + i_{s,d}\lambda_{s,d}). \tag{2.53}$$

In the same way, it is possible to calculate the T_e in Eq. (2.39) by changing the coordinates from the $\alpha\beta$ frame to the dq frame.

The apparent power $S_{s,dq}$ can be represented as

$$S_{s,dq} = \frac{3}{2}\left(\vec{v}_{s,dq} \vec{i}^*_{s,dq}\right). \tag{2.54}$$

The stator active $P_{s,dq}$ and reactive $Q_{s,dq}$ power are components of the apparent power:

$$P_{s,dq} = P_s = \mathcal{R}e(S_{s,dq}) = \frac{3}{2}\left(v_{s,d}i_{s,d} + v_{s,q}i_{s,q}\right), \tag{2.55}$$

$$Q_{s,dq} = Q_s = \mathcal{I}m(S_{s,dq}) = \frac{3}{2}\left(v_{s,q}i_{s,d} - v_{s,d}i_{s,q}\right), \tag{2.56}$$

where $\mathcal{I}m$ is the imaginary component and $\mathcal{R}e$ is the real component of the complex number.

In the same way, the apparent power $S_{r,dq}$ of the rotor can be represented as

$$S_{r,dq} = \frac{3}{2}\left(\vec{v}_{r,dq}\,\vec{i}^{\,*}_{r,dq}\right). \tag{2.57}$$

As mentioned, the rotor active $P_{r,dq}$ and reactive $Q_{r,dq}$ power are components of the apparent power represented as

$$P_{r,dq} = \mathcal{R}e(S_{r,dq}) = \frac{3}{2}\left(v_{r,d}i_{r,d} + v_{r,q}i_{r,q}\right), \tag{2.58}$$

$$Q_{r,dq} = \mathcal{I}m(S_{r,dq}) = \frac{3}{2}\left(v_{r,q}i_{r,d} - v_{r,d}i_{r,q}\right). \tag{2.59}$$

2.2.4 Speed dynamics representation

Here we have

$$JJ\frac{d\omega_{mec}}{dt} = T_e - T_L, \tag{2.60}$$

where JJ is the total moment of inertia and T_L is the load torque for operation as motor or mechanical torque when IM operates as a generator.

2.3 Three-phase power converter connected to the grid dynamic model

The three-phase power converter connected to the grid (CCG) guarantees the power flow from the electrical grid and to the DC link, or in the opposite way, operating as an inverter. When the power flows from the grid to the DC link, the converter operates as a rectifier. The converter links the electrical grid by means of a filter.

The converter is a power electronic device that makes AC–DC or DC–AC conversion by using semiconductor devices such as diode, insulated gate bipolar transistor (IGBT), metal oxide semiconductor field effect transistor (MOSFET), silicon controlled rectifier (SCR), etc. In this book, the IGBT with a diode in anti-parallel orientation will be employed, and its real characteristics, such as voltage drops and switching time, are not considered. So, it operates as an ideal switch. Fig. 2.3 depicts the components of the CCG as described, and the power electronic converter employs IGBT plus diode in anti-parallel switches.

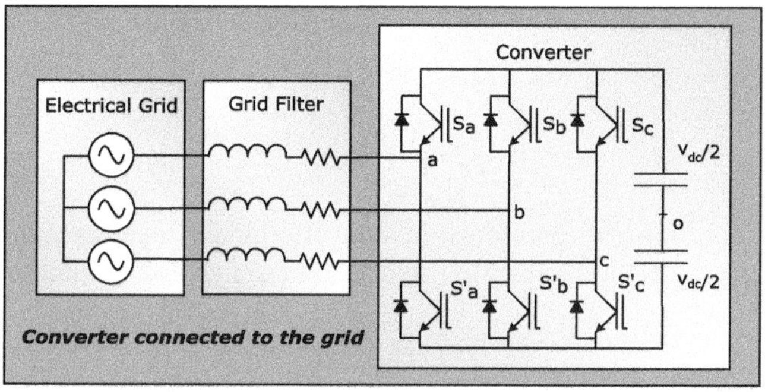

FIGURE 2.3 Detailed circuit of CCG.

The filter is allocated between each phase of the converter and the electrical grid. It can be made by using at least three inductors (L-type filter) or another arrangement by using a capacitor together with inductors as LC and LCL. The voltage of the electrical grid is fed by a transformer and it is considered sinusoidal and balanced under normal operation conditions. The effect of the grid impedance and the transformer are neglected in this chapter.

2.3.1 Three-phase power CCG representation in three-phase systems

The model of three-phase power CCG can be achieved by considering the converter depicted in Fig. 2.3 as a variable voltage source. In this case, the converter can apply voltage with several frequency and amplitude values. The simplified circuit of a phase of the CCG in this scheme is depicted in Fig. 2.4.

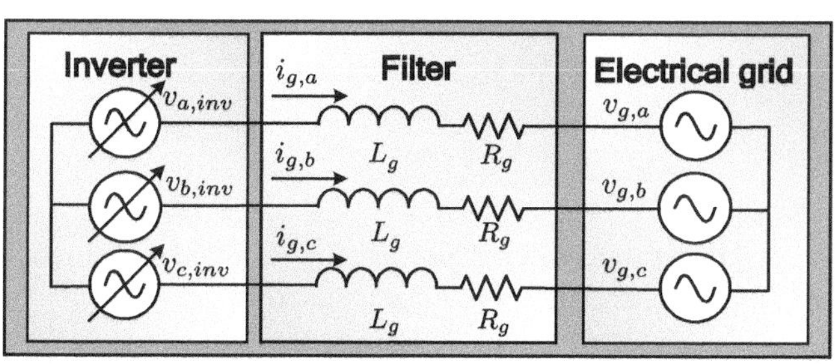

FIGURE 2.4 Simplified circuit of CCG.

The CCG in the three-phase systems, by using the scheme presented for one phase in Fig. 2.4, can be represented as:

$$v_{a,inv} = i_{g,a} R_g + L_g \frac{di_{ag}}{dt} + v_{a,g}, \tag{2.61}$$

$$v_{b,inv} = i_{g,b} R_g + L_g \frac{di_{bg}}{dt} + v_{g,b}, \tag{2.62}$$

$$v_{c,inv} = i_{g,c} R_g + L_g \frac{di_{cg}}{dt} + v_{g,c}, \tag{2.63}$$

where the subscript inv represents the converter and g represents the grid, R_g and L_g represent the inductor filter internal resistance and inductance, respectively.

2.3.2 CCG representation in stationary frame ($\alpha\beta$)

The CCG representation in the stationary frame can be achieved by using the coordinate transformation. In this case, the system model is described by means of the space vector presented in Sections 2.1 and 2.1.1 by using Eqs. (2.61)–(2.63):

$$\vec{v}_{inv,\alpha\beta} = R_g \vec{i}_{g,\alpha\beta} + L_g \frac{d\vec{i}_{g,\alpha\beta}}{dt} + \vec{v}_{g,\alpha\beta}, \tag{2.64}$$

which means

$$v_{inv,\alpha} = R_g i_{g,\alpha} + L_g \frac{di_{g,\alpha}}{dt} + v_{g,\alpha} \tag{2.65}$$

and

$$v_{inv,\beta} = R_g i_{g,\beta} + L_g \frac{di_{g,\beta}}{dt} + v_{g,\beta}, \tag{2.66}$$

where vector $\vec{v}_{inv,\alpha\beta}$ is the converter voltage, $\vec{v}_{g,\alpha\beta}$ is the grid voltage, and $\vec{i}_{g,\alpha\beta}$ is the grid current. Also, $\vec{v}_{inv,\alpha\beta} = v_{inv,\alpha} + jv_{inv,\beta}$, $\vec{i}_{g,\alpha\beta} = i_{g,\alpha} + ji_{g,\beta}$, and $\vec{v}_{g,\alpha\beta} = v_{g,\alpha} + jv_{g,\beta}$.

2.3.2.1 *Power equations*

The apparent power $S_{g,\alpha\beta}$ can be represented as

$$S_{g,\alpha\beta} = \frac{3}{2}[\vec{v}_{g,\alpha\beta} \vec{i}^{\,*}_{g,\alpha\beta}]. \tag{2.67}$$

The components of the apparent power (Eq. (2.67)) are the reactive $Q_{g,\alpha\beta}$ and active $P_{g,\alpha\beta}$ power, and they can be calculated as

$$P_{g,\alpha\beta} = P_g = \frac{3}{2}\mathcal{R}e(S_{g,\alpha\beta}) = \frac{3}{2}\left(v_{g,\alpha} i_{g,\alpha} + v_{g,\beta} i_{g,\beta}\right) \tag{2.68}$$

and

$$Q_{g,\alpha\beta} = Q_g = \frac{3}{2}\mathcal{I}m(S_{g,\alpha\beta}) = \frac{3}{2}\left(v_{g,\beta}i_{g,\alpha} - v_{g,\alpha}i_{g,\beta}\right). \tag{2.69}$$

2.3.3 CCG representation in synchronous frame *dq*

The CCG representation in the synchronous reference frame can be achieved by using the coordinate transformation. In this case, the system model is described by means of the space vector presented in Sections 2.1 and 2.1.2 by multiplying Eq. (2.64) by $e^{-j\theta_g}$ and it can be represented as

$$\vec{v}_{inv,dq} = R_g \vec{i}_{g,dq} + L_g \frac{d\vec{i}_{g,dq}}{dt} + \vec{v}_{g,dq} + j\omega_g L_g \vec{i}_{g,dq}, \tag{2.70}$$

which means

$$v_{inv,d} = R_g i_{g,d} + L_g \frac{di_{g,d}}{dt} + v_{g,d} + \omega_g L_g i_{g,q} \tag{2.71}$$

and

$$v_{inv,q} = R_g i_{g,q} + L_g \frac{di_{g,q}}{dt} + v_{g,q} - \omega_g L_g i_{g,d}. \tag{2.72}$$

In Eq. (2.70), it is necessary to note the expression below:

$$\frac{d\vec{i}_{g,\alpha\beta}}{dt} e^{-j\theta_g} = \frac{d\vec{i}_{g,dq}}{dt} + j\omega_g \vec{i}_{g,dq}, \tag{2.73}$$

where $\vec{v}_{inv,dq}$ expresses the converter voltage vector, $\vec{v}_{g,dq}$ represents the grid voltage vector, $\vec{i}_{g,dq}$ expresses the grid current vector, and ω_g represents the grid angular frequency. Also, $\vec{v}_{inv,dq} = v_{inv,d} + v_{inv,q}$, $\vec{i}_{g,dq} = i_{g,d} + i_{g,q}$, and $\vec{v}_{g,dq} = v_{g,d} + v_{g,q}$.

2.3.3.1 *Power equations*

The apparent power $S_{g,dq}$ can be expressed as follows:

$$S_{g,dq} = \frac{3}{2}[\vec{v}_{g,dq} \vec{i}^{\,*}_{g,dq}]. \tag{2.74}$$

The components of the apparent power (Eq. (2.74)) are the reactive $Q_{g,dq}$ and active $P_{g,dq}$ power and they can be calculated as

$$P_{g,dq} = P_g = \frac{3}{2}\mathcal{R}e(S_{g,dq}) = \frac{3}{2}\left(v_{g,d}i_{g,d} + v_{g,q}i_{g,q}\right) \tag{2.75}$$

and

$$Q_{g,dq} = Q_g = \frac{3}{2}\mathcal{I}m(S_{g,dq}) = \frac{3}{2}\left(v_{g,q}i_{g,d} - v_{g,d}i_{g,q}\right). \tag{2.76}$$

2.4 Pulse-width-modulation techniques

This section presents the sinusoidal, sinusoidal plus third harmonic, and space vector pulse-width-modulation (PWM) [10]. The mentioned modulation techniques can be employed to control the output voltage of the converter by means of the activation of the switches of the converter. In this way, it is possible to have voltages with several frequencies and amplitudes at the inverter output.

2.4.1 Sinusoidal PWM

This section introduces the sinusoidal PWM (SPWM) [10–12] theory for the activation of the switches of the converter, as depicted in Fig. 2.3. In this PWM technique, the carrier is a saw-tooth wave that is compared with a voltage reference signal called modulating. The result of this comparison will activate the switches on each arm of the converter so that the switches on the same arm cannot be turned on at the same time instance. In SPWM, the frequency and amplitude of the fundamental harmonic component are determined by the frequency of the reference voltage, and the carrier defines the switching frequency. A comparison of the carrier V_{por} and the modulating V_{mod} will result in the following voltage values:

$$\begin{aligned} &\text{If } V_{mod} > V_{port} \text{ then } s_x = 1 \rightarrow v_{x0} = +\frac{V_{dc}}{2}; \\ &\text{If } V_{mod} < V_{port} \text{ then } s_x = 0 \rightarrow v_{x0} = -\frac{V_{dc}}{2}, \end{aligned} \tag{2.77}$$

where $x = a, c$, or b.

The relationship between the carrier and modulating waves' amplitude values is known as the amplitude modulation index and it can be represented by

$$m = \frac{V_{mod}}{V_{port}}, \tag{2.78}$$

where $m \leq 1$ guarantees the operation in the linear range of the modulation. In this way, $v_{first-harm} = m \cdot \frac{V_{dc}}{2}$ will be the value received by the peak value of the fundamental component of the voltage.

The block diagram for the SPWM can be seen in Fig. 2.5 in which the reference voltages v_a, v_b, and v_c (modulating signals), carrier (triangular wave), and other items necessary for its implementation can be observed. So it is possible to activate the switches s_a, s_b, and s_c. It is important to mention that the signals s_a', s_b', and s_c' have the complementary value of the signals s_a, s_b, and s_c. Hence, if $s_a = 1$ then $s_a' = 0$, and vice versa.

The waveform of the carrier (saw-tooth wave), three-phase voltages, v_{a0}, v_{b0}, and v_{ab} for a three-phase SPWM are depicted in Fig. 2.6. A comparison of the carrier and the modulating signals is depicted in Fig. 2.6(a), the voltages v_{a0} (Fig. 2.6(b)) and v_{b0} (Fig. 2.6(c)) can be seen using Fig. 2.3, and the line voltage v_{ab} (Fig. 2.6(d)) will be applied to the load.

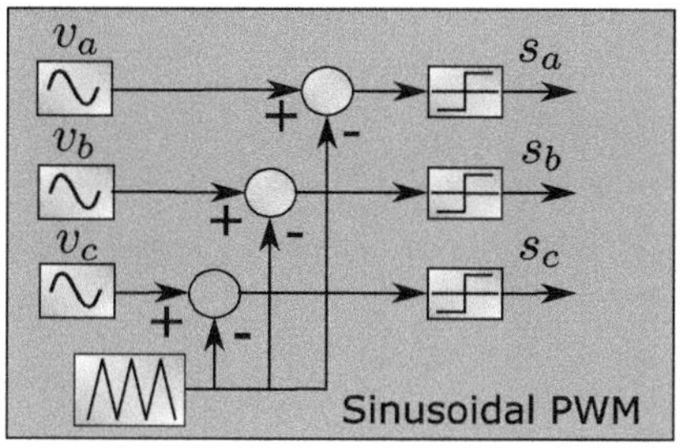

FIGURE 2.5 Block diagram for implementation of the SPWM.

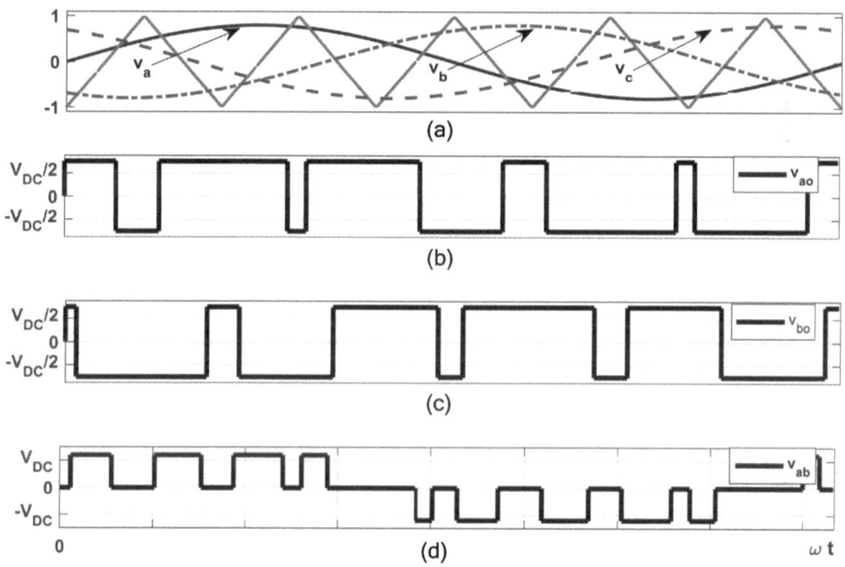

FIGURE 2.6 SPWM curves: (a) comparison of the carrier and the modulating signals, (b) v_{a0}, (c) v_{b0}, and (d) v_{ab}.

2.4.1.1 Sinusoidal PWM with third harmonic injection

The sinusoidal PWM with third harmonic injection (THSPWM) is a type of sinusoidal pulse width modulation where a harmonic component is added to the voltage signal so that the waveform of the modulating signal has its top flattened and one reduces the period of over modulation. As a consequence, the fundamental component of the voltage has greater amplitude with less harmonic distortion. In the case of this section, the third harmonic component will be employed together with the fundamental component, as shown in Fig. 2.7. However, the amplitude of this component must be less than the fundamental component. It is mentioned that, in such a case, this component cannot be seen in the output voltage. The fundamental component of the voltage is increased by 1.15 without being in the overmodulation region in a situation where the ampli-

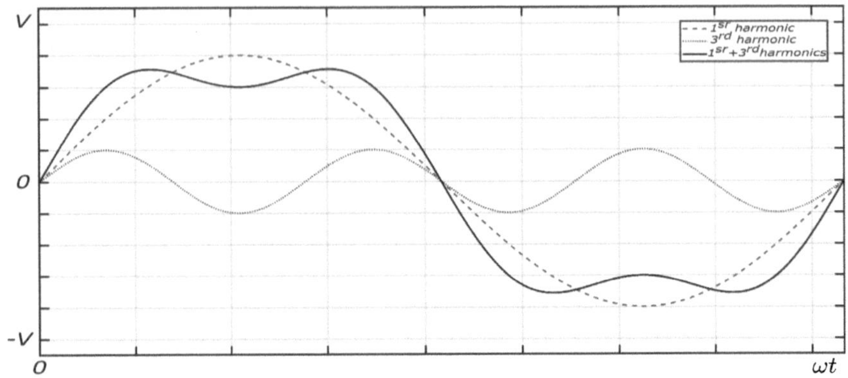

FIGURE 2.7 Voltage by using third harmonic component.

tude of the third harmonic component is 1/6 of the amplitude of the fundamental component [11,12]. The third harmonic component of the voltage can be calculated using the three-phase voltages, as presented in the following expression [10]:

$$v_{third-harmonic} = -\frac{\min\{v_a, v_b, v_c\} + \max\{v_a, v_b, v_c\}}{2}. \tag{2.79}$$

The diagram for the THSPWM is shown in Fig. 2.8. It is observed that Eq. (2.79) is employed for the calculation of the third harmonic component that is added to the three-phase voltages. Hence, the mentioned voltages will be compared with the saw-tooth wave, so it is possible to activate the switches s_a, s_b, and s_c. Again, it is important to mention that the signals s_a', s_b', and s_c' have the complementary value of the signals s_a, s_b, and s_c. Hence, if $s_a = 1$ then $s_a' = 0$, and vice versa.

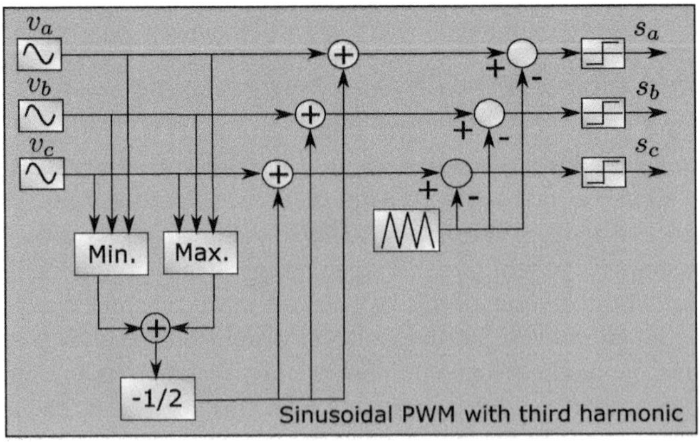

FIGURE 2.8 THSPWM diagram.

2.4.2 Space vector modulation

This section introduces steps to implement the symmetric space vector modulation (SVM) technique due to the advantages of this method over traditional methods, as better employment of the DC link [10]. In the following sections, aspects related to the scaling of input variables to the SVM algorithm, determination of the sector of the reference vector and the calculation of the times for the PWM signals for digital implementation will be discussed, as presented in [13].

2.4.2.1 Scaling of input signals to the SVM algorithm

The SVM algorithm receives the components of the references $v_{\alpha_{ref}}$ and $v_{\beta_{ref}}$ from the desired voltage vector in the $\alpha\beta$ frame and the voltage value of the DC bus of the inverter V_{DC}. At the output of the algorithm, three PWM signals are obtained to activate the inverter switches. The values of $v_{\alpha_{ref}}$ and $v_{\beta_{ref}}$ enter the algorithm in normalized format using $\frac{\sqrt{3}}{V_{DC}}$ due to the maximum possible voltage value as a function of the bus voltage value.

Fig. 2.9 and Table 2.1 show a representation and values of the fixed voltage vectors for the different combinations of the inverter switch states. The area between two consecutive vectors in the plan is called a sector and, therefore, there are six distinct sectors. The reference vector $\vec{v}_{ref}$ can be expressed using a linear association of the base vectors that limit its sector (adjacent vectors) and the null vectors. In Fig. 2.9, the base vectors are named V_1, V_2, V_3, V_4, V_5, and V_6, the null vectors that do not appear in the figure are $V_0 = (000)$ and $V_7 = (111)$.

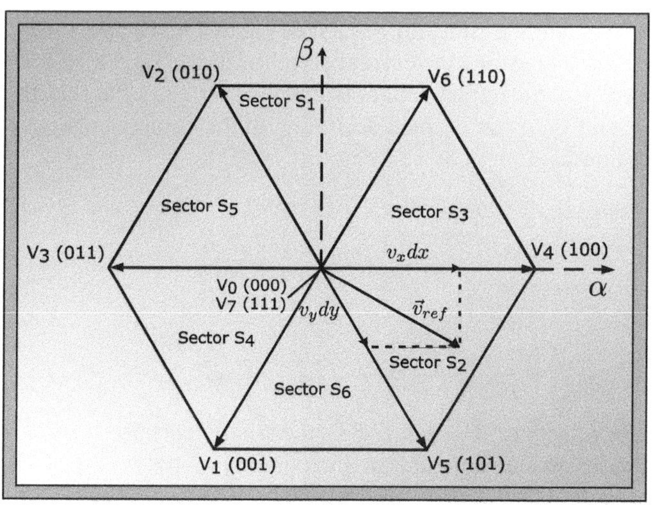

FIGURE 2.9 Space vector representation by using the switches states in the $\alpha\beta$ frame.

TABLE 2.1 Switching states and voltage vectors.

S_a	S_b	S_c	Voltage vector
0	0	0	$V_0 = 0$
1	0	0	$V_1 = \frac{2}{3}V_{dc}$
1	1	0	$V_2 = \frac{1}{3}V_{dc} + j\frac{\sqrt{3}}{3}V_{dc}$
0	1	0	$V_3 = -\frac{1}{3}V_{dc} + j\frac{\sqrt{3}}{3}V_{dc}$
0	1	1	$V_4 = -\frac{2}{3}V_{dc}$
0	0	1	$V_5 = -\frac{1}{3}V_{dc} - j\frac{\sqrt{3}}{3}V_{dc}$
1	0	1	$V_6 = \frac{1}{3}V_{dc} - j\frac{\sqrt{3}}{3}V_{dc}$
1	1	1	$V_7 = 0$

The reference vector $\vec{v}_{ref}$ can be decomposed according to the base vectors of the sector and the null vectors by

$$\vec{v}_{ref} = dxv_x + dyv_y + dzv_z, \tag{2.80}$$

where v_x and v_y are the limit vectors of the sector where the reference vector is located, v_z indicates the null vector applied, also $dx = \frac{t_1}{t_{PWM}}$, $dy = \frac{t_2}{t_{PWM}}$, and $dz = \frac{t_{PWM}-t_1-t_2}{t_{PWM}}$ are the time fractions related to the PWM period that the vectors v_x, v_y, and v_z remain applied in order to synthesize the original reference vector.

2.4.2.2 Determination of the reference voltage vector sector

In order to calculate the times of the PWM signals using the SVM algorithm, the first step is to determine the current sector of the reference vector $\vec{v}_{ref}$. One way to accomplish this implementation is reported in [13]. For this, the variables v_{1ref}, v_{2ref}, and v_{3ref} are defined according to the components $v_{\alpha_{ref}}$ and $v_{\beta_{ref}}$ from expressions:

$$v_{1ref} = v_{\beta_{ref}}, \tag{2.81}$$

$$v_{2ref} = \frac{1}{2}(\sqrt{3}v_{\alpha_{ref}} - v_{\beta_{ref}}), \tag{2.82}$$

and

$$v_{3ref} = \frac{1}{2}(-\sqrt{3}v_{\alpha_{ref}} - v_{\beta_{ref}}). \tag{2.83}$$

From Eqs. (2.81), (2.82), and (2.83), the coefficients B, A, and C are calculated, which can take a value of 0 or 1 depending on the v_{1ref}, v_{2ref}, and v_{3ref} sign, respectively. For this, the following rules apply:

- **If** $v_{1ref} > 0$ **then** $A = 1$, **else** $A = 0$;
- **If** $v_{1ref} > 0$ **then** $B = 1$, **else** $B = 0$;

- **If** $v_{1ref} > 0$ **then** $C = 1$, **else** $C = 0$.

With the values obtained for the coefficients A, B, and C, the sector of the reference vector is calculated by

$$sector = 2B + A + 4C. \tag{2.84}$$

2.4.2.3 Determination of times for PWM signals

Once the sector of the reference vector has been determined, the second step consists of calculating the times t_1 and t_2 using the base vectors that limit the sector. Table 2.2 shows the definition of the times t_1 and t_2 for each sector of the $\alpha\beta$ frame, where the variables X_1 and Y_1 are defined for a period of the PWM signal T_{PWM} using the following expressions [13]:

$$X_{SVM1} = \sqrt{3}\frac{T_{PWM}}{V_{DC}} v_{1\beta_{ref}}, \tag{2.85}$$

$$Y_{SVM1} = \left(\frac{\sqrt{3}}{2} v_{1\beta_{ref}} + \frac{\sqrt{3}}{2} v_{1\alpha_{ref}}\right)\frac{T_{PWM}}{V_{DC}}, \tag{2.86}$$

and

$$Z_{SVM1} = \left(\frac{\sqrt{3}}{2} v_{1\beta_{ref}} - \frac{\sqrt{3}}{2} v_{1\alpha_{ref}}\right)\frac{T_{PWM}}{V_{DC}}. \tag{2.87}$$

TABLE 2.2 PWM time scale definition.

Time/sector	1	2	3	4	5	6
t_1	Z_{SVM1}	Y_{SVM1}	$-Z_{SVM1}$	$-X_{SVM1}$	X_{SVM1}	$-Y_{SVM1}$
t_2	Y_{SVM1}	$-X_{SVM1}$	X_{SVM1}	Z_{SVM1}	$-Y_{SVM1}$	$-Z_{SVM1}$

If, during the execution of the algorithm, the sum of the times t_1 and t_2 is greater than the period of PWM, the algorithm establishes the saturation condition, in this way, the expressions used are given by

$$t_{1sat} = t_1\frac{T_{PWM}}{t_1 + t_2} \tag{2.88}$$

and

$$t_{2sat} = t_2\frac{T_{PWM}}{t_1 + t_2}. \tag{2.89}$$

In order to obtain symmetric PWM signals from the SVM algorithm, it is necessary to calculate the times t_a, t_b, and t_c of these signals. This is done using the following expression [13]:

$$t_{svm-a} = \frac{T_{PWM} - t_1 - t_2}{2}, \tag{2.90}$$

$$t_{svm-b} = t_{svm-a} + t_1, \tag{2.91}$$

and

$$t_{svm-c} = t_{svm-b} + t_2. \tag{2.92}$$

The values of t_{svm-a}, t_{svm-b}, and t_{svm-c} are then expressed in units of the register of DSP accounts and loaded into the DSP comparison registers CMPR1, CMPR2, and CMPR3. The values loaded in the mentioned registers depend on the current sector of the reference vector, as shown in Table 2.3. Once this procedure has been carried out, the DSP generates at its output the six PWM's signals necessary to control the inverter.

TABLE 2.3 Times for registers in function of the sector.

Registers/sector	1	2	3	4	5	6
PWM_1	t_{svm-b}	t_{svm-b}	t_{svm-a}	t_{svm-c}	t_{svm-c}	t_{svm-b}
PWM_2	t_{svm-a}	t_{svm-c}	t_{svm-b}	t_{svm-b}	t_{svm-a}	t_{svm-c}
PWM_3	t_{svm-b}	t_{svm-b}	t_{svm-c}	t_{svm-a}	t_{svm-b}	t_{svm-a}

Fig. 2.10 depicts the SVM digital implementation using the vectors V_4 and V_5 of sector 2, as depicted in Fig. 2.9. It is possible to see the vector sequence that is active and their duration times.

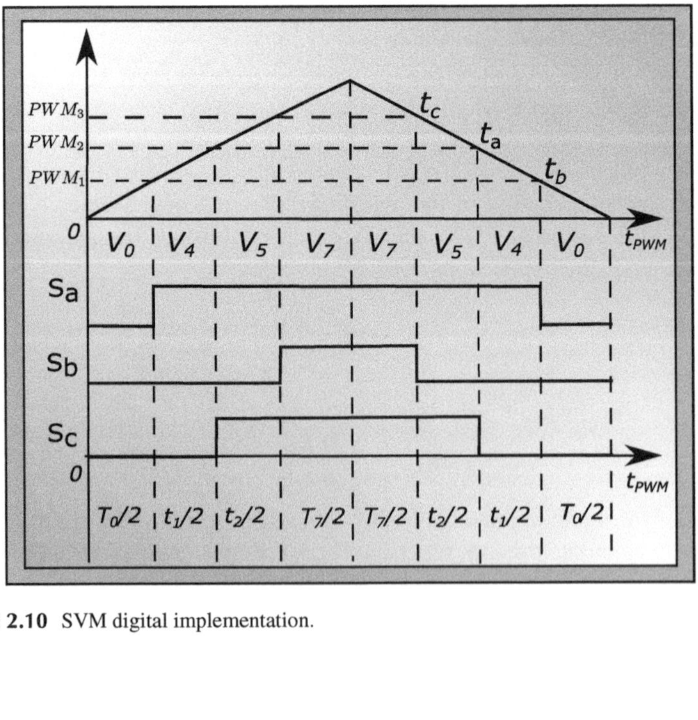

FIGURE 2.10 SVM digital implementation.

2.5 Summary

This chapter introduces the space vector representation, the coordinates transformations ($\alpha\beta \rightarrow dq$ and $dq \rightarrow \alpha\beta$), the IM and three-phase power CCG dynamic models representation in the abc, $\alpha\beta$, and dq frames, and finally, the algorithms to the SPWM, THSPWM, and SVM. Additionally, the expressions of the electromagnetic torque, the reactive and the active power, and diagrams are also described. These pieces of information are the base to the power or current control implementation in the next chapters.

2.6 Further reading

The steady state and dynamics characteristics of the IM are discussed in [9,14–23].

Regarding the power electronics converters, modulation and control are presented in [10–12,24–27].

Chapter 3

Fundamentals of vector control for DFIG and for the three-phase CCG

Alfeu J. Sguarezi Filho
Federal University of ABC, Santo André, SP, Brazil

3.1 Doubly-fed induction generator

The doubly-fed induction generator (DFIG) is a type of IM that operates as generator and it can be supplied at terminals of the stator and the rotor. Hence, the rotor and the stator have windings. Due to concerns about CO_2 emissions, the wind energy stations [3], small hydroturbines [28], and the tidal [29] energy plants have interesting applications of the DFIG.

Regarding the scheme, Fig. 3.1 depicts the DFIG connected to the electrical grid. In this case, the stator is connected directly to the grid. The rotor of DFIG is connected to the mains by means of a back-to-back power electronic converter. This converter is composed by two DC–AC converters that share the same DC link. In this configuration, the power electronic converter processes 25% of the DFIG nominal power due its stability. Additionally, the cost of the converter decreases.

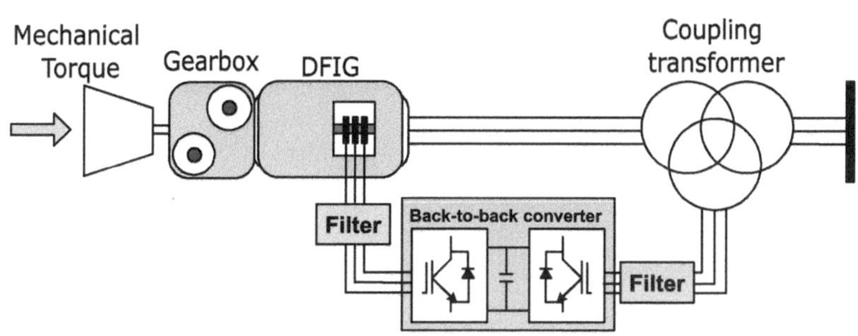

FIGURE 3.1 DFIG connected to the mains block diagram.

Table 3.1 presents the operation points of the doubly-fed induction machine when one operates as a motor (1 and 4) and as a generator (2 and 3) according

https://doi.org/10.1016/B978-0-32-390964-8.00012-9

to the speed, slip, mechanical power, and electromagnetic torque. In generator operation mode, at supersynchronous speed ($\omega_r > \omega_s$), the power flows from the rotor of DFIG to the DC link and to the electrical grid by means of the CCG. At subsynchronous speed ($\omega_r < \omega_s$), the power flows from the DC link to the rotor of DFIG.

TABLE 3.1 Doubly-fed induction machine operation conditions.

Operation mode	T_e	Speed	Slip	Power		
				P_m	P_s	P_r
1 Motor	$T_e > 0$	$\omega_r > \omega_s$	$slip < 0$	$P_m > 0$	$P_s > 0$	$P_r > 0$
2 Generator	$T_e < 0$	$\omega_r > \omega_s$	$slip < 0$	$P_m < 0$	$P_s < 0$	$P_r < 0$
3 Generator	$T_e < 0$	$\omega_r < \omega_s$	$slip > 0$	$P_m < 0$	$P_s < 0$	$P_r > 0$
4 Motor	$T_e > 0$	$\omega_r < \omega_s$	$slip > 0$	$P_m > 0$	$P_s > 0$	$P_r < 0$

P_m mechanical power, P_s stator power, and P_r rotor power.

3.1.1 Vector control

In this scheme the synchronous reference frame (dq) rotates with an angular frequency equal to the rotation velocity of the stator flux vector, defined by

$$\frac{d\theta_s}{dt} = \omega_s . \tag{3.1}$$

Hence, the stator flux space vector $\vec{\lambda}_{s,\alpha\beta}$ in the dq frame can be represented as

$$\begin{aligned} \vec{\lambda}_{s,dq} &= \lambda_{s,d} + j\lambda_{s,q} = \vec{\lambda}_{s,\alpha\beta}\, e^{-j\theta_s} \\ &= \left[\lambda_{s,\alpha} + j\lambda_{s,\beta}\right]\left[\cos(\theta_s) - j\sin(\theta_s)\right]. \end{aligned} \tag{3.2}$$

In this way, the components of the stator flux space vector $\vec{\lambda}_{s,dq}$ are:

$$\begin{aligned} \lambda_{s,d} &= \left[\lambda_{s,\alpha}\cos(\theta_s) + \lambda_{s,\beta}\sin(\theta_s)\right], \\ \lambda_{s,q} &= \left[-\lambda_{s,\alpha}\sin(\theta_s) + \lambda_{s,\beta}\cos(\theta_s)\right]. \end{aligned} \tag{3.3}$$

In the stator oriented system, the components of the stator flux vector become

$$\lambda_s = \lambda_{s,d} = \left|\vec{\lambda}_{s,dq}\right| = \sqrt{\lambda_{s,d}^2 + \lambda_{s,q}^2} \tag{3.4}$$

and

$$\lambda_{s,q} = 0. \tag{3.5}$$

By using Eqs. (3.4) and (3.5), the relationship between the current and fluxes represented by Eqs. (2.50) and (2.51) respectively becomes:

$$\lambda_s = L_s i_{s,d} + L_M i_{r,d} \tag{3.6}$$

and

$$0 = L_s i_{s,q} + L_m i_{r,q}. \tag{3.7}$$

So,

$$\lambda_{r,d} = L_m i_{s,d} + L_r i_{r,d} \tag{3.8}$$

and

$$\lambda_{r,q} = L_m i_{s,q} + L_r i_{r,q}. \tag{3.9}$$

The elements of the stator voltage space vector ($\vec{v}_{s,dq}$) can be expressed as follows:

$$\begin{aligned}\vec{v}_{s,dq} &= v_{s,d} + j v_{s,q} = \vec{v}_{s,\alpha\beta}\, e^{-j\theta_s} \\ &= \left[v_{s,\alpha} + j v_{s,\beta}\right]\left[\cos(\theta_s) - j\sin(\theta_s)\right],\end{aligned} \tag{3.10}$$

which means

$$\begin{aligned}v_{s,d} &= \left[v_{s,\alpha}\cos(\theta_s) + v_{s,\beta}\sin(\theta_s)\right], \\ v_{s,q} &= \left[-v_{s,\alpha}\sin(\theta_s) + v_{s,\beta}\cos(\theta_s)\right].\end{aligned} \tag{3.11}$$

The stator voltage components using stator flux orientation in steady state actions using Eq. (2.46) can be represented as

$$v_{s,d} = 0 \tag{3.12}$$

and

$$v_{s,q} = v_s = \left|\vec{v}_{s,dq}\right| \cong \omega_s \lambda_s. \tag{3.13}$$

It can be noticed in Eqs. (3.12) and (3.13) that, when the stator flux is oriented along the direct axis, the voltage aligns with the quadrature axis. In this way, the stator resistance R_s is neglected in Eq. (2.46) due to the fact that the resistance of the stator is very small when IM has more than some hundreds of kW and the element $\frac{d\vec{\lambda}_{s,dq}}{dt} = 0$ due to assumed steady state operation.

The expression of the stator active power (2.55) can be represented using the rotor current by using Eqs. (3.7), (3.12), and (3.13) as

$$P_s = -\frac{3}{2} v_s \frac{L_m}{L_s} i_{r,q}. \tag{3.14}$$

In the same way, the expression of the stator reactive power (2.56) can be represented by means of the rotor current by using Eqs. (3.6), (3.12), and (3.13) as

$$Q_s = \frac{3}{2} v_s \left(\frac{\lambda_s}{L_s} - \frac{L_m}{L_s} i_{r,d}\right). \tag{3.15}$$

It can be noticed that when the DFIG operates under normal grid voltage conditions, the stator active (3.14) and reactive (3.15) power can be manipulated by means of the rotor current control using the converter connected to the rotor and the stator of the DFIG connected to the grid.

3.1.1.1 Frame orientations

The fundamental concepts of vector control are achieved by the choice of the angle to make the axis transformations. Eqs. (3.12) and (3.13) show that when the stator flux is oriented along the direct axis, the voltage aligns with the quadrature axis. In this way, there are two possibilities to align the axis:

- Align the estimated stator flux along the direct (d) axis, stator flux orientation;
- Align the stator/grid voltage vector along the quadrature (q) axis, voltage orientation.

Initially, the DFIG control employed the stator flux orientation. Under normal grid operation conditions, the stator flux and the stator voltage are stable. Hence, the stator flux vector is shifted by 90° relative to the stator voltage vector when the stator resistance is neglected. Due to grid disturbance such as voltage sag and harmonic distortions, the phase-locked loop (PLL) can compute the position of the voltage vector and it is shifted by 90° to make the stator flux and voltage vector have a coincident reference frame. Hence, the stator voltage vector reaches the alignment in the q axis, the direct d axis is shifted by 90°, and the stator voltage is the same as the electrical grid voltage. In this way, in this book the choice is to align the stator voltage vector along the quadrature q axis. For simplicity, it will be called stator voltage oriented control (SVOC). Thus, we can use the relationship presented in the last section and in the references [3] and [30] which can give more details about this type of orientation. It can be noticed that, when the stator of DFIG is linked to the main grid, the stator and grid voltage are the same vector. The voltage orientation (Fig. 3.2(a)) and the stator flux orientation (Fig. 3.2(b)) are depicted in Fig. 3.2.

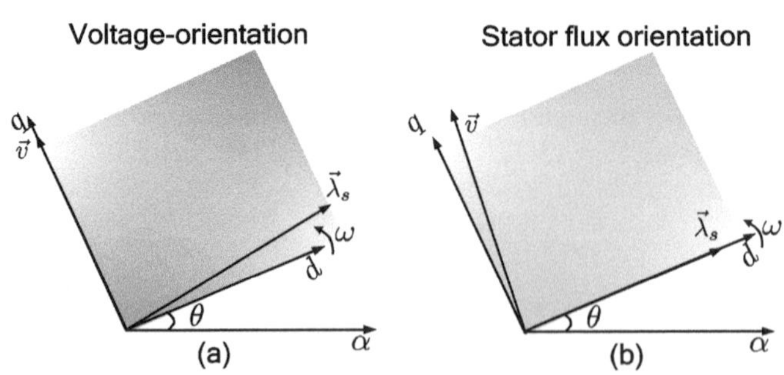

FIGURE 3.2 Orientation diagrams: (a) voltage orientation and (b) stator flux orientation.

3.1.1.2 Transformations employed in this book

This section introduces the transformations employed in this book and all coordinate transformations are described in Sections 2.1.1 and 2.1.2. In this way, the frame transformations can be expressed as follows:

- Rotor frame – from abc to the $\alpha\beta r$ frame, one employs Eq. (2.3);
- Stationary frame – from abc to the $\alpha\beta$ frame, one employs Eq. (2.3);
- dq frame for DFIG – from $\alpha\beta r$ to the dq frame, one employs Eq. (2.7) using $\theta = \theta_{sl} = \theta_s - \theta_r$;
- dq frame for DFIG – from $\alpha\beta$ to the dq frame, one employs Eq. (2.7) using $\theta = \theta_s$
- dq frame for CCG: from $\alpha\beta$ to the dq frame, one employs Eq. (2.7) using $\theta = \theta_g$.

In the opposite direction to that presented, only use the inverse transformations (see Sections 2.1.1 and 2.1.2).

3.1.2 Closed loop rotor current control using PI controllers

The closed loop rotor current control permits the reactive and active stator power control separately, as presented in Eqs. (3.14) and (3.15), by using SVOC. Initially, the three-phase rotor current is measured by sensors and they are transformed from the rotor reference frame $\alpha\beta r$ by using the Clarke's transform, and to the synchronous reference frame (dq) by using the slip angle ($\theta_{sl} = \theta_s - \theta_r$) and the rotor current vector. Hence, two PIs (proportional-integral) controllers process the error between the rotor current vector ($\vec{i}_{r,dq} = i_{r,d} + j i_{r,q}$) elements and their references so that they calculate the rotor voltage vector ($\vec{v}_{r,dq} = v_{r,d} + j v_{r,q}$) elements in the synchronous reference frame (dq), as presented in Eq. (3.18). The PWM algorithm generates switching signals for the rotor side converter using rotor voltage vector $\vec{v}_{r,\alpha\beta r} = \vec{v}_{r,dq} e^{j(\theta_s - \theta_r)}$ in the rotor reference frame employing the slip angle ($\theta_{sl} = \theta_s - \theta_r$).

The angle θ_s is estimated by means of the stator voltage measured by sensors and a PLL. The main component of the DC link is the capacitors. The power injected to the DC link is in accordance with the DFIG speed operation, as presented in Table 3.1. Fig. 3.3 depicts DFIG rotor current control diagram by using the PIs and SVOC. The stator power references can be represented as

$$i_{r,qref} = -\frac{2 P_{ref} L_s}{3 v_s L_m} \tag{3.16}$$

and

$$i_{r,dref} = -\frac{2 Q_{ref} L_s}{3 v_s L_m} + \frac{\lambda_s}{L_m}, \tag{3.17}$$

where P_{ref} and Q_{ref} represent the stator active and reactive power references, respectively.

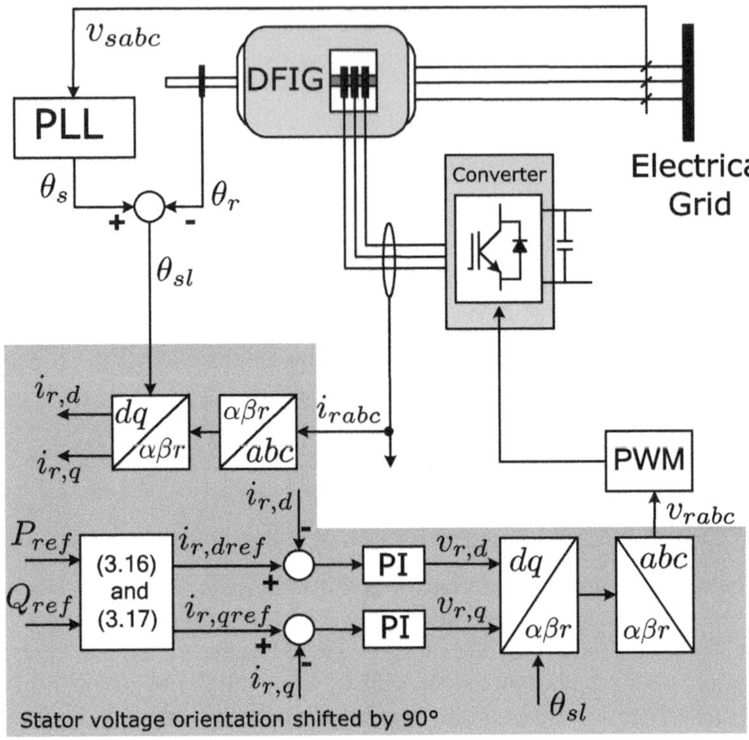

FIGURE 3.3 DFIG rotor current control diagram.

The components of the rotor voltage vector computed by the PI controller can be written as follows:

$$
\begin{aligned}
v_{r,d} &= (i_{r,dref} - i_{r,d}) \cdot PI, \\
v_{r,q} &= (i_{r,qref} - i_{r,q}) \cdot PI,
\end{aligned} \tag{3.18}
$$

where the PI controller can be expressed as

$$
PI = k_p + \frac{k_i}{s}. \tag{3.19}
$$

The proportional and integral gains of the PI controller can be calculated for a settling time of 10 ms as [31]:

$$
\begin{aligned}
k_p &= \frac{1}{5 \cdot 10^{-3}} \sigma L_r, \\
k_i &= \frac{1}{5 \cdot 10^{-3}} L_r.
\end{aligned} \tag{3.20}
$$

It is possible to use the same values of the gains in the two control loops.

3.1.3 Deadbeat rotor current control for DFIG

The deadbeat technique introduces a control theory that permits computing an input in which one guarantees a null steady state error using the discrete representation of the system [32]. In this way, the mentioned control strategy was employed to obtain the rotor voltage vector using discrete equations of the DFIG as introduced in [33,34]. Hence, the technique permits null rotor current error in the dq frame and a fast settling time.

In this way, the rotor voltage vector can be represented as just current and voltage vectors by substituting Eq. (2.51) into Eq. (2.48) as

$$\begin{aligned}\vec{v}_{r,dq} &= R_r \vec{i}_{r,dq} + \frac{d}{dt}\left(L_m \vec{i}_{s,dq} + L_r \vec{i}_{r,dq}\right) \\ &+ j\left(\omega_s - \omega_r\right)\left(L_m \vec{i}_{s,dq} + L_r \vec{i}_{r,dq}\right),\end{aligned} \tag{3.21}$$

which means

$$v_{r,d} = R_r i_{r,d} + \frac{d}{dt}\left(L_m i_{s,d} + L_r i_{r,d}\right) - \left(\omega_s - \omega_r\right)\left(L_m i_{s,q} + L_r i_{r,q}\right) \tag{3.22}$$

and

$$v_{r,q} = R_r i_{r,q} + \frac{d}{dt}\left(L_m i_{s,q} + L_r i_{r,q}\right) + \left(\omega_s - \omega_r\right)\left(L_m i_{s,d} + L_r i_{r,d}\right). \tag{3.23}$$

The rotor voltage vector is represented by means of the stator flux magnitude and rotor current vector. The representation is achieved by substituting the stator current in Eq. (3.6) into Eq. (3.22) for the direct axis and, similarly, substituting Eq. (3.7) into Eq. (3.23) for the quadrature axis. Considering the DFIG bonded to the main grid and under normal operation conditions, we get the relationship $\frac{d\lambda_1}{dt} = 0$. Hence, the rotor voltage vector components become:

$$v_{r,d} = R_r i_{r,d} - L_r \omega_{sl} i_{r,q} + \left(\frac{L_s L_r - L_m^2}{L_s}\right)\frac{di_{r,d}}{dt} + \omega_{sl}\frac{L_m^2}{L_s} i_{r,q} \tag{3.24}$$

and

$$v_{r,q} = R_r i_{r,q} + L_r \omega_{sl} i_{r,d} + \left(\frac{L_s L_r - L_m^2}{L_s}\right)\frac{di_{r,q}}{dt} + \omega_{sl}\left(\frac{L_m}{L_s}\lambda_s - \frac{L_m^2}{L_s} i_{r,d}\right). \tag{3.25}$$

Considering $\sigma = 1 - \frac{L_m^2}{L_s L_r}$ and $\sigma L_r = \frac{L_s L_r - L_m^2}{L_s}$, Eqs. (3.24) and (3.25) become

$$v_{r,d} = R_r i_{r,d} - \omega_{sl} \sigma L_r i_{r,q} + \sigma L_r \frac{di_{r,d}}{dt} \tag{3.26}$$

and

$$v_{r,q} = R_r i_{r,q} + \omega_{sl} \sigma L_r i_{r,d} + \sigma L_r \frac{di_{r,q}}{dt} + \omega_{sl}\frac{L_m}{L_s}\lambda_s, \tag{3.27}$$

which means

$$\vec{v}_{r,dq} = (R_r + j\sigma L_r\omega_{sl})\,\vec{i}_{r,dq} + \sigma L_r\frac{d\vec{i}_{r,dq}}{dt} + j\omega_{sl}\frac{L_m}{L_s}\lambda_s, \tag{3.28}$$

where $\omega_{sl} = \omega_s - \omega_r$.

The discrete equation can be reached by means of Euler's forward method [5,32],

$$\frac{dx(t)}{dt} \cong \frac{x(k+1) - x(k)}{T}, \tag{3.29}$$

where $x(k+1)$ is the next sample, $x(k)$ is the actual sample, and T is the sampling time.

The discretized rotor voltage components can be reached by employing Eq. (3.29) in Eqs. (3.26) and (3.27). To calculate the rotor vector by using the deadbeat control technique, it is assumed that the samples at $(k+1)$th instant are the references, i.e., $i_{r,q}(k+1) = i_{r,qref}$ and $i_{r,d}(k+1) = i_{r,dref}$. Hence, the elements that constitute the rotor voltage vector can be represented by:

$$v_{r,d}(k) = R_r i_{r,d}(k) - \omega_{sl}(k)\sigma L_r, i_{r,q}(k) + \sigma L_r\frac{i_{r,dref} - i_{r,d}(k)}{T} \tag{3.30}$$

and

$$\begin{aligned} v_{r,q}(k) &= R_r i_{r,q}(k) + \omega_{sl}(k)\sigma L_r i_{r,d}(k) \\ &\quad + \sigma L_r\frac{i_{r,qref} - i_{r,q}(k)}{T} + \omega_{sl}(k)\frac{L_m}{L_s}\lambda_s(k), \end{aligned} \tag{3.31}$$

which means

$$\begin{aligned} \vec{v}_{r,dq}(k) &= (R_r + j\sigma L_r\omega_{sl}(k))\,\vec{i}_{r,dq}(k) \\ &\quad + \sigma L_r\frac{\vec{i}_{r,dq}(k+1) - \vec{i}_{r,dq}(k)}{T} + j\omega_{sl}(k)\frac{L_m}{L_s}\lambda_s(k), \end{aligned} \tag{3.32}$$

where $\lambda_s(k) = \frac{v_{s,q}(k)}{\omega_s(k)}$ as presented in Eq. (3.13).

The rotor current references can be represented by means of the stator reactive and active power. In this way, $i_{2q_{ref}}$ can be presented by using Eq. (3.14):

$$i_{r,q}(k+1) = i_{r,qref} = -\frac{2P_{ref}L_s}{3v_s L_m}, \tag{3.33}$$

where P_{ref} acts as the reference signal of the stator active power.

In the same way, $i_{r,d_{ref}}$ can be presented by using Eq. (3.15):

$$i_{r,d}(k+1) = i_{r,dref} = -\frac{2Q_{ref}L_s}{3v_s L_m} + \frac{\lambda_s(k)}{L_m}, \tag{3.34}$$

where Q_{ref} acts as the reference signal of the stator reactive power.

The rotor voltage vector components are as in Eqs. (3.30) and (3.31) in the dq frame. Thus, the rotor voltage vector ($\vec{v}_{r,\alpha\beta r}$) in the $\alpha\beta r$ frame can be calculated by using $\vec{v}_{r,\alpha\beta r} = \vec{v}_{r,dq} e^{j(\theta_s - \theta_r)}$, as presented in Eqs. (2.10)–(2.12). The vector $\vec{v}_{r,\alpha\beta r}$ can be the input of the PWM technique. Hence, the constituent elements of the rotor current vector will achieve the reference signals.

The DFIG deadbeat rotor current control diagram is represented by Fig. 3.4 and the details of the deadbeat diagram are depicted in Fig. 3.5.

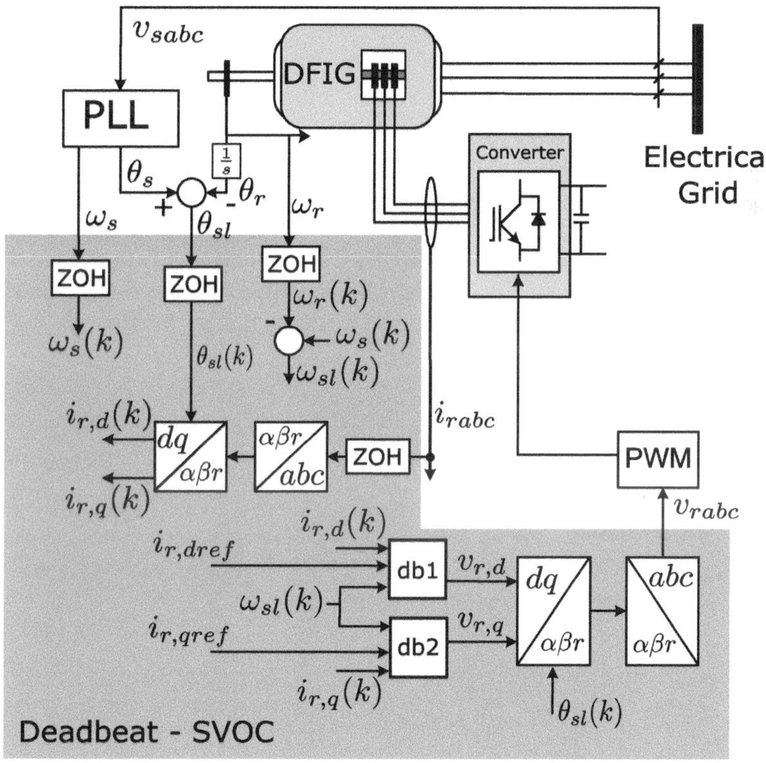

FIGURE 3.4 DFIG deadbeat rotor current control diagram.

3.1.4 Deadbeat direct power control for DFIG

In the same way as introduced in Section 3.1.3, this section is about the deadbeat direct power control for DFIG.

3.1.4.1 Fundamentals of direct power control

The direct power control (DPC) concept is accomplished without current closed-loops and it handles directly the stator reactive Q_s and active P_s power of DFIG. In this way, the mentioned technique performs the rotor flux of the generator

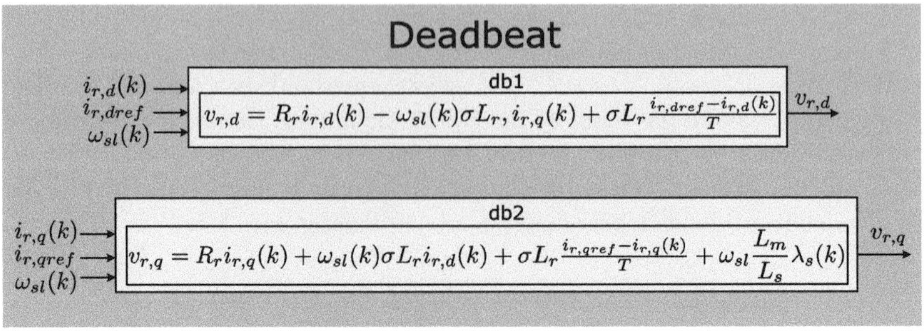

FIGURE 3.5 Detail of deadbeat diagram.

control by means of the rotor voltage vector. Hence, the stator reactive and active powers are expressed using the constituent elements of the rotor flux vector and SVOC.

In this way, it is possible to employ Eqs. (3.14) and (3.15) to calculate the stator power. The relationship between the constitutive elements of the rotor flux and current vectors can be represented by manipulating the Eqs. (3.7)–(3.9) presented as

$$i_{r,d} = \frac{L_s}{L_s L_r - L_m^2}\lambda_{r,d} - \frac{L_m}{L_s}\frac{L_s}{L_s L_r - L_m^2}\lambda_s, \quad (3.35)$$

$$i_{r,q} = \frac{L_s}{L_s L_r - L_m^2}\lambda_{r,q}. \quad (3.36)$$

The active (Eq. (3.14)) and reactive (Eq. (3.15)) powers can be figured out as per Eqs. (3.35) and (3.36), and they become:

$$P_s = -\frac{3 v_s L_M}{2\sigma L_s L_r}\lambda_{r,q}, \quad (3.37)$$

$$Q_s = \frac{3 v_s L_M}{2\sigma L_s L_r}\left(-\lambda_{r,d} + \frac{L_r}{L_M}\lambda_s\right). \quad (3.38)$$

When the rotor resistance is neglected, the rotor flux vector is directly manipulated by the rotor voltage vector, as represented by Eq. (2.48), during a small time interval. Thus, the stator power can be manipulated as per the rotor flux represented by Eqs. (3.37) and (3.38). This is the fundamental concept to the DPC of stator active and reactive power using the rotor side converter when the DFIG has its stator linked directly to the grid.

3.1.4.2 Rotor side equations for DPC

In the DPC technique, the DFIG has to be expressed using the rotor voltage vector, the stator reactive and the active power. In this way, the constitutive elements of the rotor voltage vector in Eq. (2.48) can be represented using the SVOC and the active (3.37) and the reactive (3.38) power, neglecting the rotor resistance as follows:

$$\frac{dQ_s}{dt} = \frac{v_{r,d}}{Am} + \omega_{sl} P_s \tag{3.39}$$

and

$$\frac{dP_s}{dt} = \frac{v_{r,q}}{Am} - \omega_{sl} Q_s - \omega_{sl} \frac{L_r}{L_m\, Am} \lambda_s, \tag{3.40}$$

where $\omega_{sl} = \omega_s - \omega_r$, $\sigma = 1 - \frac{L_m^2}{L_s L_r}$ is the leakage coefficient, and $Am = -\frac{2\sigma L_s L_r}{3 v_s L_m}$.

3.1.4.3 DPC using deadbeat control for DFIG

The deadbeat controller is used by the DPC representation to obtain rotor voltages, as presented in Section 3.1.3. The deadbeat control using the discrete state equations of the DFIG was introduced in [35,36]. In this way, the mentioned DPC deadbead strategy for DFIG was employed to obtain the rotor voltage vector using discrete equations of the DFIG by means of the stator power errors. Hence, the technique permits null stator reactive and active power errors and a fast settling time.

The discrete rotor voltage elements are computed as per Eq. (3.29) in Eqs. (3.39) and (3.40). Thus, they can be represented as follows:

$$v_{r,d} = Am \frac{Q_{ref} - Q_s(k)}{T} - \omega_{sl}(k) P_s(k) Am \tag{3.41}$$

and

$$v_{r,q} = Am \frac{P_{ref} - P_s(k)}{T} + \omega_{sl}(k) Q_s(k) Am + \omega_{sl}(k) \frac{L_r \lambda_s}{L_m T}, \tag{3.42}$$

where $\lambda_s = \frac{v_{s,q}(k)}{\omega_s(k)}$ is as presented in Eq. (3.13), $P_{ref} = P_s(k+1)$, and $Q_{ref} = Q_s(k+1)$.

The rotor voltage vector elements are in accordance to Eqs. (3.41) and (3.42) in the dq frame. Hence, the rotor voltage vector ($\vec{v}_{r,\alpha\beta r}$) in the $\alpha\beta r$ reference frame can be calculated by using $\vec{v}_{r,\alpha\beta r} = \vec{v}_{r,dq} e^{j(\theta_s - \theta_r)}$, as presented in Eqs. (2.10)–(2.12). The vector $\vec{v}_{r,\alpha\beta r}$ can be the input of the PWM technique. Thus, the elements of the apparent power will reach the references of the stator reactive and active power. The deadbeat DPC for DFIG scheme is represented by Fig. 3.6.

3.2 Three-phase power CCG vector control

This section introduces the fundamentals of the CCG current control using the vector control technique.

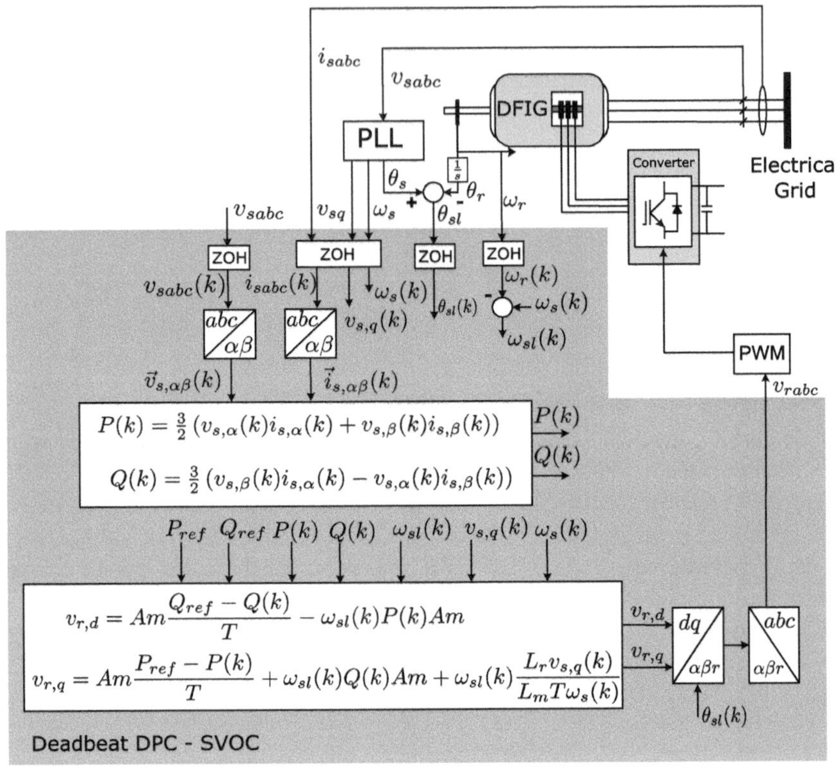

FIGURE 3.6 DFIG deadbeat direct power control general scheme.

3.2.1 Filter elements

In this book the filter elements are the capacitor to the DC link, C, and the inductor to link the CCG to the electrical grid, L_g. The cited elements can be expressed as follows [37–40]:

$$L_g = \frac{v_{dc}}{\Delta I \cdot freq},$$
$$C = \frac{P}{v_{dc}\Delta v_{dc} \cdot freq}, \tag{3.43}$$

where ΔI and Δv_{dc} are the desired current and voltage ripples, respectively, and $freq$ is the switching frequency. It is interesting to evidence that the cited works can employ gains in Eq. (3.43).

3.2.2 Vector control fundamentals for the CCG

The CCG is linked to the electrical grid employing a filter. In this way, in this book the choice is to align the converter voltage vector along the quadrature d

axis. So, it can be called the grid voltage orientated control (GVOC). The CCG model can be seen in Section 2.3.3.

The angular frequency of the dq frame rotates with velocity equal to the rotation velocity of the grid voltage vector, defined by

$$\frac{d\theta_g}{dt} = \omega_g. \tag{3.44}$$

The values of θ_g and ω_g are estimated by a PLL that employs the three-phase grid voltages.

The elements of the grid voltage vector $\vec{v}_{g,dq}$ can be represented as

$$\begin{aligned}\vec{v}_{g,dq} &= v_{g,d} + j v_{g,q} = \vec{v}_{g,\alpha\beta}\, e^{-j\theta_g} \\ &= \left[v_{g,\alpha} + j v_{g,\beta}\right]\left[\cos(\theta_g) - j\sin(\theta_g)\right],\end{aligned} \tag{3.45}$$

which means

$$\begin{aligned}v_{g,d} &= \left[v_{g,\alpha}\cos(\theta_g) + v_{g,\beta}\sin(\theta_g)\right], \\ v_{g,q} &= \left[-v_{g,\alpha}\sin(\theta_g) + v_{g,\beta}\cos(\theta_g)\right].\end{aligned} \tag{3.46}$$

When the GVOC is employed, the components of the voltage vector can be represented as

$$v_{g,d} = v_g = \left|\vec{v}_{g,dq}\right| \tag{3.47}$$

and

$$v_{g,q} = 0. \tag{3.48}$$

Substituting Eqs. (3.47) and (3.48) into Eqs. (2.75) and (2.76) ($P_{g,dq} = P_g = \frac{3}{2}\left(v_{g,d}i_{g,d} + v_{g,q}i_{g,q}\right)$ and $Q_{g,dq} = Q_g = \frac{3}{2}\left(v_{g,q}i_{g,d} - v_{g,d}i_{g,q}\right)$, respectively, the power can be represented as

$$P_{g,dq} = P_g = \frac{3}{2}\left(v_{g,d}i_{g,d}\right) \tag{3.49}$$

and

$$Q_{g,dq} = Q_g = \frac{-3}{2}\left(v_{g,d}i_{g,q}\right). \tag{3.50}$$

It can be noticed that, when the CCG operates under normal grid voltage conditions, the grid active (3.49) and reactive (3.50) power can be manipulated employing the grid current control by using the grid side converter.

3.2.2.1 *Closed loop grid current control by using PI controllers*

The closed loop of the grid current control allows the reactive and active stator power control separately, as presented in Eqs. (3.49) and (3.50) by using GVOC.

Initially, the three-phase grid current is measured by sensors and they are transformed to the $\alpha\beta$ frame by using Clarke's transform, and to the synchronous reference frame (dq), by using the grid angle (θ_g) and the grid current vector. The angle θ_g is estimated by means of the electrical grid voltage measured by sensors and the PLL.

Therefore, two PI (proportional-integral) controllers process the error using the grid current vector ($\vec{i}_{g,dq} = i_{g,d} + j i_{g,q}$) and their references so that they calculate the converter voltage vector ($\vec{v}_{inv,dq} = v_{inv,d} + j v_{inv,q}$) in the synchronous reference frame (dq), as presented in Eq. (3.51). The PWM algorithm enables the signals that drive the grid side converter using the elements of the grid voltage vector in the stationary reference frame employing the grid angle ($\vec{v}_{inv,\alpha\beta} = \vec{v}_{inv,dq} e^{j\theta_g}$):

$$\begin{aligned} v_{inv,d} &= (i_{g,dref} - i_{g,d}) \cdot PI - \omega_g L_g i_{g,q}, \\ v_{inv,q} &= (i_{g,qref} - i_{g,q}) \cdot PI + \omega_g L_g i_{g,d}, \end{aligned} \tag{3.51}$$

where $PI = k_p + \frac{k_i}{s}$.

The DC link voltage is manipulated by means of the direct axis component of the current grid vector ($i_{g,d}$). In this case, the PI processes the error between the DC link voltage reference and the measured DC link voltage. One calculates the reference of the direct axis component of the grid current vector ($i_{g,dref}$).

Fig. 3.7 depicts a three-phase power CCG current control diagram by using the PI controllers and GVOC. The reference of the quadrature axis component of the current grid vector ($i_{g,qref}$) is manipulated by means of the reactive power reference and it can be represented isolating ($i_{g,q}$) in Eq. (3.50).

3.3 Summary

This chapter introduces the concepts of the vector control to DFIG and to CCG, the frame orientations, the rotor current control of the DFIG using the deadbeat control theory and PI controllers, the DPC concepts, and DPC for DFIG using deadbeat control theory. The deadbeat employs the discrete representation of the DFIG. Additionally, the grid current control using PI controllers for the CCG is introduced.

In this way, the described techniques permit controlling each element of the current vector – or the elements of the apparent power – independently using the vector control principles and the PWM.

3.4 Further reading

The references [2,24–27,41–43] present information about the power electronics converters, modulation, and control.

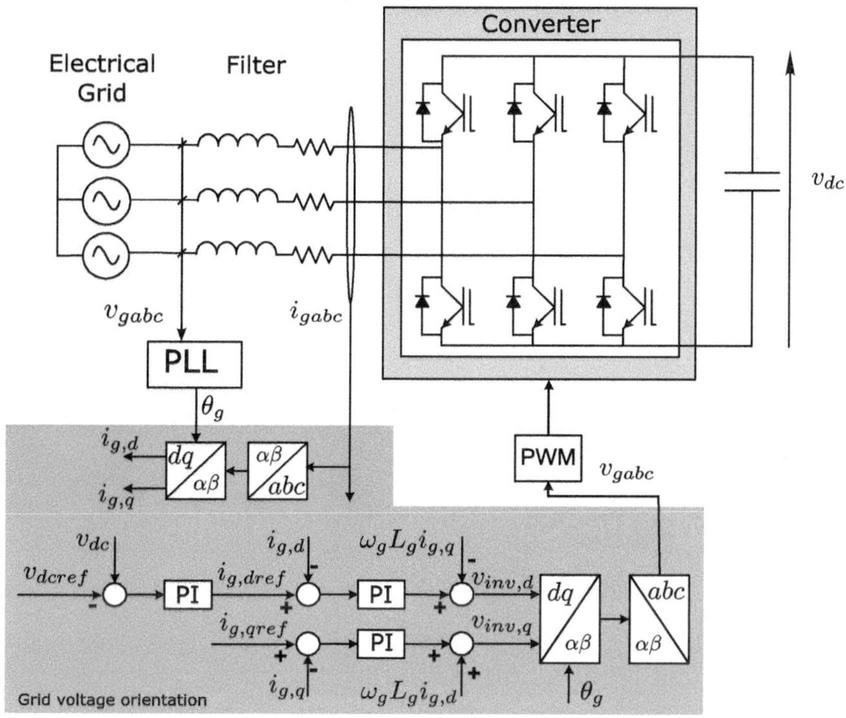

FIGURE 3.7 Three-phase power CCG closed-loop grid current control by using PI controllers.

Regarding the IM drives, the references [2,3,15,17,44–46] present several types of oriented control techniques, estimation, etc.

Chapter 4

Fundamentals of model predictive control

Alfeu J. Sguarezi Filho
Federal University of ABC, Santo André, SP, Brazil

4.1 Overview

Model predictive control (MPC) is a control technique that has been studied since the 1970s [4,47] and encompasses a range of controllers that use this term. In this sense, MPC employs the model of the plant in order to calculate its input, minimizing a cost/objective function. MPC has the following advantages [47,48]: it can be used in processes with many variables; restrictions can be easily introduced into the control process during its development; it is an easy and useful application, when the references to be reached are fixed or previously known. As for disadvantages, the derivation of the control law is more complex when compared with that of classic controllers, such as PI; the process time increases considerably when considering the constraints, and, finally, the performance of the controller can be affected depending on the way the plant is modeled.

In this context, the basic idea of the MPC can be concatenated as follows:

- It permits predicting the future behavior of the plant at points in time;
- Calculation of the input signal to be applied is done with the minimization of a cost function;
- The input signal is updated every time.

The basic scheme of the MPC can be seen in Fig. 4.1. In this scheme, the plant model is employed to predict the future behavior of the plant's outputs through the values, together with the future values calculated for the input. It is noted that the optimizer is responsible for calculating the inputs or controlling actions from the cost function that use the errors between the reference and future outputs and that have restrictions. The system can be modeled using state equations, transfer function, impulse or step response, so that its representation has a good precision [4]. Regarding the optimizer, it provides control actions based on the calculation of inputs using the minimized cost function. This value can be obtained by means of the signs of future reference points, signs of exits and past entries. This does not happen with classic control techniques that calculate the input so the output reaches the reference according to past errors.

Model Predictive Control for Doubly-Fed Induction Generators and Three-Phase Power Converters
https://doi.org/10.1016/B978-0-32-390964-8.00013-0

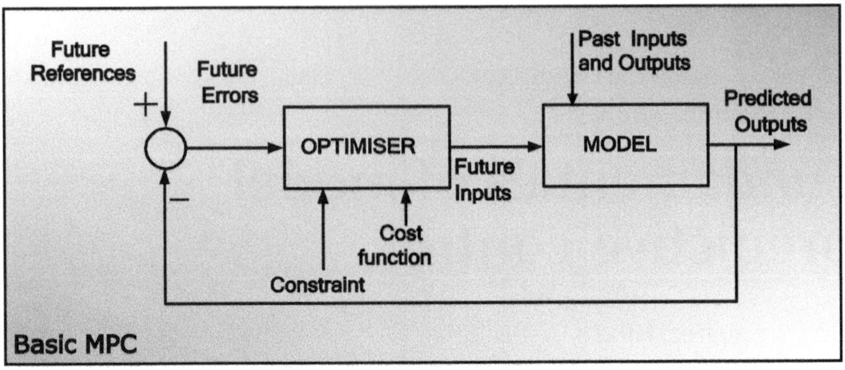

FIGURE 4.1 MPC basic diagram.

Most MPCs employ the concept of free and forced responses when calculating inputs or control actions. In this case, the objective is to represent the control sequence by adding the free and forced response, respectively, as introduced in [47]

$$u(t) = u_c(t) + u_f(t), \tag{4.1}$$

where $u_c(t)$ is the next value calculated by the controller in the future considering the past values null and $u_f(t)$ represents the past values of the entry kept constant and identical to the value at the last value of the variable manipulated in the future time instants. In this way, the evolution of the system's response in relation to its current state corresponds to the free response, whereas the forced response takes into account future control actions.

The type of objective cost function used also represents differences between the predictive controllers. The cost function employed, which considers the quadratic error, can be represented by

$$J_c = \sum_{i_{mpc}=n_1}^{i_{mpc}=n_y} \Xi_e \left(r_{i_{mpc}} - x_{i_{mpc}}\right)^2 + \sum_{i_{mpc}=1}^{i_{mpc}=n_u} \Xi_{ui} \Delta u_{i_{mpc}}^2, \tag{4.2}$$

where n_1 represents the initial horizon of prediction of the exits, n_y is the final horizon of prediction of the exits, n_u represents the horizon of control, Ξ_{ui} is the weight in the control signal, Ξ_e is the weighting of errors, $x_{i_{mpc}}$ is the prediction of the outputs steps forward, i.e., the controlled variable calculated at each sampling instant k, $r_{i_{mpc}}$ is the future reference signal, and $\Delta u_{i_{mpc}}$ is the variation in the control signal, that is, variation in the manipulated variable calculated at k.

The control effort corresponding to the second term of Eq. (4.2) may not be considered in some methods, while in others the values of the control signal may be represented directly and not as increments. In addition to the quadratic error in the cost function, the expression in discrete and continuous states, respectively, can be expressed as follows:

$$J_c = |x_{ref} - x(k+1)|,$$
$$J_c = \int_{k+1}^{k} (x(t) - x(t)_p)dt, \tag{4.3}$$

where $|\cdot|$ is the absolute value and $_p$ is the predicted value of the state.

4.1.1 MPC applied in power electronics systems

Due to the evolution of digital signal processors and the use of MPC in systems that permits a quick (in milliseconds) response, the applications in power electronics become possible in the 1980s in high power systems [49]. There are several studies that used some type of MPC applied to the power electronics and some will be mentioned to show the use of this control technique. The generalized predictive control was applied to the control of the induction motor [50,51] or to the control of a single-phase inverter [52], MPC was employed in the control of the current of a voltage source inverter [53], the functional predictive control was applied to DFIG [54], predictive direct control was applied to an active filter [55], and the finite control set was applied to various types of power electronic drives [56]. MPC has also been applied to the DC/DC converters for photovoltaic systems [57], and direct predictive torque control has been applied to the permanent magnet motor in [58].

4.2 Finite control set model predictive control

4.2.1 Principles of finite control set model predictive control

The finite control set MPC (FCS-MPC) is a type of predictive control that uses the system model in its discrete form and its minimized cost function defines the state to be used depending on the possible states of the converter [48,56]. In this sense, the FCS-MPC operates with a variable switching frequency and without the use of PWM techniques, which makes it simpler from the point of view of implementing its algorithm.

In the FCS-MPC technique, the prediction of the future behavior of the system is performed using the discretized state equation from Eq. (3.29). This representation in general can be done as follows [4]:

$$x(k+1) = Ax(k) + Bu(k), \tag{4.4}$$

where $u(k)$ is the system input, $x(k)$ expresses the current state at the sampling time T, $x(k+1)$ is the future sample, A is the state matrix, and B is the input matrix. In some cases, the disturbance in the expression presented in Eq. (4.4) can be considered.

From the point of view of the cost function, it can be formulated according to the errors from the power P and Q, current vector $\vec{i}$, voltage vector $\vec{v}$ or torque T_e, and λ fluxes. These expressions are given by

$$\begin{aligned}
J_c &= |P_{ref} - P(k+1)| + |Q_{ref} - Q(k+1)| \text{ or} \\
J_c &= |\vec{i}_{ref} - \vec{i}(k+1)| \text{ or} \\
J_c &= |\vec{v}_{ref} - \vec{v}(k+1)| \text{ or} \\
J_c &= |T_{eref} - T_e(k+1)| + |\lambda_{ref} - \lambda(k+1)|,
\end{aligned} \tag{4.5}$$

where $|\cdot|$ is the absolute value, $_{ref}$ is the reference and $(k+1)$ is the prediction of each variable in an instant of sampling and one can be greater than one sampling time.

It can be seen that Eq. (4.5) employs only the left part of the cost function presented in Eq. (4.2) for choosing the vector.

From the discrete form of the system model, the possible states of the converter are employed. In this chapter, a two-level inverter will be used in which the states are shown in Table 2.1. So, this FCS-MPC technique calculates the future behavior of the plant using each possible mentioned state. These values are compared with the desired reference, and the chosen state is that which allows the least error.

Algorithm 4.1 FCS-MPC algorithm

while MPC-FCS active **do**
  The system variables such as current, reactive and active power or flux and torque are measured
  $J_{op} = \infty$
  for
  h = 0 : 7 **do**
    $\vec{v}(h) =$ function of V_{dc} % Possible states of the converter
    $\vec{x}(k+1) = A\vec{x}(k) + B\vec{v}(h)$ % System prediction of the state x
    $J_c(h) = |x_{ref} - x(k+1)|$ % Cost function calculation
    if $J_c(h) < J_{op}$ **then**
      $J_{op} = J_c(h)$
      $h_{op} = h$ % Store the state with the least cost
    end if
    if $h = 7$ **then**
      Apply $\vec{v}(h_{op})$
      Start again
      Else go to **for** loop again
    end if
  end for
end while

FCS-MPC pseudocode without constraints can be seen in Algorithm 4.1 for the case of FCS-MPC using only a future sample prediction. In this technique, the computational cost increases exponentially when the prediction horizon increases due to the fact that, for each future prediction, all states of the converter are employed in this task.

4.2.2 Constrains in FCS-MPC

An important feature of MPC is the possibility of using constrains in the formulation of the controller and, consequently, the design of this component can achieve a better compromise between the reference and the control effort. In power electronics systems or electrical machine drives, the current, switching frequency, switching losses, or voltage are related to the control effort. In this sense, it is enough to add an element in the cost function in a way the control effort can be weighted. The cost function expression in which the constraint is considered can be represented as

$$J_c = |x_{ref} - x(k+1)| + \Lambda|\Delta|, \tag{4.6}$$

where Λ is the weighting constant, and Δ is related to the constraint or the penalty of the desired variation or maximum limit, which may be speed or voltage, power, switching frequency, torque, fluxes or current [56,59–62]. In the case of a voltage variation $\Delta = v(k-1) - v(k)$, and depending on the value chosen for Λ, the same voltage value will be kept applied for more sampling periods.

4.2.3 Modulated finite control set model predictive control

The FCS-MPC technique operates at a variable switching frequency with an average value between $\frac{1}{3}$ and $\frac{1}{4}$ of the sampling frequency [48,56]. Thus, the operation of this controller with fixed switching frequency was initially a challenge for works presented in [63–65] and this proposal was called modulated MPC. In this book it will be called modulated FCS-MPC.

The modulated FCS-MPC performs the prediction of the variables' behavior according to the same procedure employed for FCS-MPC (Eq. (4.4)) and in each sampling period the technique assesses the prediction of two active vectors using two cost functions. The cost function for each active vector operates in the same way as in FCS-MPC. To illustrate this operation, if the vector is in sector 3 (Fig. 2.9), the first cost prediction and function evaluates vector $\vec{v}_4$ and the second prediction and cost function evaluates vector $\vec{v}_6$. The duty cycles can be calculated using the equations [66]:

$$d_0 = \frac{K_j}{J_0},$$
$$d_1 = \frac{K_j}{J_1}, \tag{4.7}$$

$$d_2 = \frac{K_j}{J_2},$$
$$d_1 + d_2 + d_0 = 0,$$

where d_0 is the duty cycle of the null vector to be applied. From the solution of Eq. (4.7), it is possible to calculate K_j and the values of the work cycles as given by [66]:

$$\begin{aligned} d_0 &= \frac{T J_1 J_2}{(J_0 J_1 + J_1 J_2 + J_0 J_2)}, \\ d_1 &= \frac{T J_0 J_2}{(J_0 J_1 + J_1 J_2 + J_0 J_2)}, \\ d_2 &= \frac{T J_0 J_1}{(J_0 J_1 + J_1 J_2 + J_0 J_2)}. \end{aligned} \tag{4.8}$$

Using Eqs. (4.7) and (4.8), a new cost function equation can be obtained with the use of duty cycles for the active vectors, namely

$$J_c(k+1) = J_1 d_1 + J_2 d_2. \tag{4.9}$$

4.3 MPC with modulator (MPC-WM)

Another way to design the MPC from Section 4.1 is to use it in conjunction with a PWM as described in Section 2.4. In this type of application, MPC calculates the voltage vector to be applied instead of using the possible states of the converter, as done in the basic scheme of FCS-MPC. The use of PWM in conjunction with MPC makes it possible to operate with fixed switching frequency more naturally.

A system can be expressed by a state equation given the following equation [4]:

$$\begin{aligned} \frac{dx_x}{dt} &= A_a x_x + B_b u_u + G_g d_d, \\ y_y &= C_c x_x, \end{aligned} \tag{4.10}$$

where u_u is the system input, x_x is the state, d_d is the disturbance, y_y is the output, A_a is the state matrix, G_g is the disturbance matrix, C_c is a output matrix, and B_b is the input matrix.

From Eq. (4.10), its discretized equation can be represented by

$$\begin{aligned} x_x(k+1) &= A_{ad} x_x(k) + B_{bd} u_u(k) + G_{gd} d_d(k), \\ y_y(k) &= C_{cd} x_x(k), \end{aligned} \tag{4.11}$$

where k is the sample at the current instant in the sampling time T, $(k+1)$ is the sample at the next instant in the state at the same sampling time, A_{ad}

is the discretized state matrix, C_{cd} is the discretized output matrix, G_{gd} is the discretized disturbance matrix, and B_{bd} is the discretized input matrix. These matrices can be expressed using Eq. (3.29) or using the equations as follows [32]:

$$\begin{aligned} A_{ad} &= e^{A_a T} \cong I + A_a T, \\ B_{bd} &= \int_0^T e^{A_a \tau} B_b \, d\tau \cong B_b T, \\ G_{gd} &= \int_0^T e^{A_a \tau} G_g \, d\tau \cong G_d T. \end{aligned} \tag{4.12}$$

From Eq. (4.11), one can find the representation for the instant $(k+2)$ as follows:

$$\begin{aligned} x_x(k+2) &= A_{ad} x_x(k+1) + B_{bd} u_u(k+1) + G_{gd} d_d(k+1), \\ y_y(k+1) &= C_{cd} x_x(k+1). \end{aligned} \tag{4.13}$$

Using Eq. (4.11), one can compute Eq. (4.13) depending on the k sample:

$$\begin{aligned} x_x(k+2) &= A_{ad}[A_{ad} x_x(k) + B_{bd} u_u(k) + G_{gd} d_d(k)] \\ &\quad + B_{bd} u_u(k+1) + G_{gd} d_d(k+1), \\ y_y(k+1) &= C_{cd}[A_{ad} x_x(k) + B_{bd} u_u(k) + G_{gd} d_d(k)]. \end{aligned} \tag{4.14}$$

Rearranging Eq. (4.14) and considering C_{cd} as an identity matrix, it is possible to obtain:

$$\begin{aligned} x_x(k+2) &= A_{ad}^2 x_x(k) + A_{ad} B_{bd} u_u(k) \\ &\quad + A_{ad} G_{gd} d_d(k) + B_{bd} u_u(k+1) + G_{gd} d_d(k+1), \\ y_y(k+1) &= A_{ad} x_x(k) + B_{bd} u_u(k) + G_{gd} d_d(k). \end{aligned} \tag{4.15}$$

In this way, $y_y(k+2)$ can be represented using Eqs. (4.14) and (4.15) as follows:

$$\begin{aligned} y_y(k+2) &= A_{ad}^2 x_x(k) + A_{ad} B_{bd} u_u(k) \\ &\quad + A_{ad} G_{gd} d_d(k) + B_{bd} u_u(k+1) + G_{gd} d_d(k+1). \end{aligned} \tag{4.16}$$

This hypothesis will help in simplifying the calculations and it makes possible to apply the above ideas in power electronics systems without great difficulty. However, if C_{cd} is not an identy matrix you can rearrange from (4.14) or see Section 3.2 in [4]. Using this same principle, $y_y(k+3), y_y(k+4), \ldots, y_y(k+n_y)$ can be represented. Therefore, the representation using the equation of state of future outputs can be written as [4,67]

$$YY = AA x_x(k) + BB\, VV + GG\, DD. \tag{4.17}$$

The predictions are expressed as $VV \in \mathbb{R}^{n_u \cdot out \times 1}$ to the inputs, $YY \in \mathbb{R}^{n_y \cdot out \times 1}$ to the outputs, $DD \in \mathbb{R}^{n_y \cdot out \times 1}$ to the disturbances, and out is the number of outputs of the system:

$$YY = \begin{bmatrix} y_y(k+1) & y_y(k+2) & y_y(k+3) & \dots & y_y(k+n_y) \end{bmatrix}^T, \tag{4.18}$$

$$VV = \begin{bmatrix} u_u(k) & u_u(k+1) & u_u(k+2) & \dots & u_u(k+n_u-1) \end{bmatrix}^T, \tag{4.19}$$

$$\begin{aligned} DD &= \begin{bmatrix} d_d(k) & d_d(k+1) & d_d(k+2) & \cdots & d_d(k+n_y-1) \end{bmatrix}^T \\ &\approx d_d(k) \begin{bmatrix} 1 & 1 & 1 & \dots & 1 \end{bmatrix}^T. \end{aligned} \tag{4.20}$$

Also $AA \in \mathbb{R}^{n_y \cdot out \times out}$, $BB \in \mathbb{R}^{n_y \cdot out \times n_u \cdot out}$, and $GG \in \mathbb{R}^{n_y \cdot out \times n_y \cdot out}$ express the matrixes in Eqs. (4.21), (4.22), and (4.23), respectively:

$$AA = \begin{bmatrix} A_{ad} & A_{ad}^2 & \dots & A_{ad}^{n_y} \end{bmatrix}^T, \tag{4.21}$$

$$BB = \begin{bmatrix} B_{bd} & 0 & 0 & \dots \\ A_{ad} B_{bd} & B_{bd} & 0 & \dots \\ A_{ad}^2 B_{bd} & A_{ad} B_{bd} & B_{bd} & \dots \\ \vdots & \vdots & \vdots & \ddots \\ A_{ad}^{n_y-1} B_{bd} & A_{ad}^{n_y-2} B_{bd} & A_{ad}^{n_y-3} B_{bd} & \dots \end{bmatrix}, \tag{4.22}$$

$$GG = \begin{bmatrix} G_{gd} & 0 & 0 & \dots & 0 \\ A_{ad} G_{gd} & G_{gd} & 0 & \dots & 0 \\ A_{ad}^2 G_{gd} & A_d G_{gd} & G_{gd} & \dots & 0 \\ \vdots & \vdots & \vdots & \ddots & \vdots \\ A_{ad}^{n_y-1} G_{gd} & A_{ad}^{n_y-2} G_{gd} & A_{ad}^{n_y-3} G_{gd} & \dots & G_{gd} \end{bmatrix}. \tag{4.23}$$

In Eqs. (4.18)–(4.23), n_y is the prediction horizon which represents the number of predictions of the outputs from the current time k, and n_u is the control horizon, that is, the number of predicted inputs to be included in the prediction model.

When the MPC operates, $n_u = 1$, the control action magnitudes represent average values and the outputs reach the references. Otherwise, the outputs reach the references more closely. A big n_u increases the computational cost.

The cost function presented in Eq. (4.2) employs the control effort and the future errors. Consequently, the future errors are calculated by taking the difference between its reference to the output $y_{yref}(k+i)$ and the prediction of the output $y_y(k+i)$, as it can be represented by:

$$\Xi(k+i) = y_y(k+i) - y_{yref}(k+i), \quad \forall\, i \in 1, 2, \dots, n_y. \tag{4.24}$$

The cost function of using quadratic future errors applying Eq. (4.24) can be represented by

$$J_c = \sum_{i=1}^{n_y} \Xi(k+i)^T W_y \Xi(k+i) + \sum_{ii=0}^{n_u-1} VV(k+j)^T W_u VV(k+j), \quad (4.25)$$

where W_y and W_u are weighting matrices.

Eq. (4.25) can be represented in the matrix form using Eqs. (4.17) and (4.24). So, it has the following representation:

$$J_c = (YY - YY_{ref})^T W_{yy}(YY - YY_{ref}) + VV^T W_{uu} VV, \quad (4.26)$$

where $W_{yy} \in \mathbb{R}^{n_y \cdot out \times n_y \cdot out}$ and $W_{uu} \in \mathbb{R}^{n_u \cdot out \times n_u \cdot out}$ are the repetitions of the weighting matrices on the main diagonal (Eqs. (4.28) and (4.29)), and YY_{ref} represents the set of future references, which can be maintained constant during the prediction period.

The control output signs can be found by the minimization of Eq. (4.26) and letting $\frac{\partial J_c}{\partial VV} = 0$ as follows:

$$VV = (BB^T W_{yy} BB + W_{uu})^{-1} BB^T W_{yy}(YY_{ref} - AAx_x - GG\ DD), \quad (4.27)$$

where W_{uu} and W_{yy} are control effort and the prediction effort weighting matrices, respectively, that are adjusted in order to obtain the desired system response. These matrices are usually diagonal and can be represented by

$$W_{yy} = \begin{bmatrix} W_{yy,1} & 0 & \cdots & 0 \\ 0 & W_{yy,2} & \cdots & 0 \\ \vdots & \vdots & \ddots & W_{yy,n_y} \end{bmatrix}, \quad (4.28)$$

$$W_{uu} = \begin{bmatrix} W_{uu,0} & 0 & \cdots & 0 \\ 0 & W_{uu,1} & \cdots & 0 \\ \vdots & \vdots & \ddots & W_{uu,n_u-1} \end{bmatrix}. \quad (4.29)$$

The elements of W_{uu} (the control effort matrix) have to be greater than zero to decrease the overshoot in the response of the system. Otherwise, is it assumed that the actuator has infinity energy. Matrix W_{yy} can improve the settling time due to the emphasis on each predicted output.

In this book, CCG and DFIG are controlled by means of the voltage vectors that are computed at each sampling time by MPC. When the voltage vector is computed by means of the MPC-WM, Eq. (4.27) is used for this task. Hence, the voltage vector will be the input of the PWM algorithm of the three-phase power converter connected to the electrical grid or to the rotor of DFIG. Due to the mentioned facts, the term MPC-WM is employed in this book.

4.3.1 Constrains in MPC

As introduced in Section 4.2.2, the constraints can be incorporated into the controller formulation. Initially, the insertion of constraints may be an expression that can be obtained using Eq. (4.17) in Eq. (4.26) so that the cost function does not depend on YY:

$$\begin{aligned} J = {} & VV^T(BB^T W_{yy} BB + W_{uu})VV \\ & + \left(2BB^T W_{yy}(AAy(k) + GGDD - YY_{ref})\right)^T VV. \end{aligned} \tag{4.30}$$

The variables in Eq. (4.30), namely $y(k)$, YY_{ref}, and DD, can be determined through measurements, or in the case of YY_{ref} from the reference value $y_{ref}(k)$, with the exception of VV. Thus, a control signal that guarantees the best response can be found by minimizing Eq. (4.26).

In this way, Eq. (4.30) can be employed in the optimization problem as presented in [68]:

$$\begin{aligned} & \underset{VV}{\text{minimize}} \quad J = \frac{1}{2} VV^T HH VV + CC^T VV \\ & \text{subject to } VV \cdot \vartheta \leq \varpi, \end{aligned} \tag{4.31}$$

where J is a quadratic function of VV whose matrices are defined as follows:

$$\begin{aligned} HH &= 2(BB^T W_{yy} BB + W_{uu}), \\ CC &= 2BB^T W_{yy}(AA\mathbf{y}(k) + GGDD - Y_{ref}). \end{aligned} \tag{4.32}$$

In this problem, the computational burden is increased when compared to FCS-MPC and is to be done offline. In addition, restrictions can be placed on outputs or inputs. The situation in which the calculated input has constraints can be represented by:

$$\begin{aligned} & \underset{VV}{\text{minimize}} \quad J = \frac{1}{2} VV^T HH VV + CC^T VV \\ & \text{subject to} \quad r = \frac{1}{2} \begin{bmatrix} u_u(k)^T u_u(k) \\ \vdots \\ u_u(k+n_u-1)^T u_u(k+n_u-1) \end{bmatrix} \leq 0. \end{aligned} \tag{4.33}$$

4.4 Summary

This chapter describes the fundamentals of MPC. This technique is a group of controllers that predict the system action and compute the input using the minimized cost function at each sampling time. Hence, one can reach the references, and the MPCs introduced are: FCS-MPC, modulated FCS-MPC, and

MPC-WM. Also, the representation of the types of cost function and the restrictions are described. In this way, the described predictive techniques permit designing MPCs to DFIG and CCG.

4.5 Further reading

Additional information about MPC theory applied in machine drives and power converters can be seen in [69–71], and for wind energy systems in [2].

Regarding the methods for solving the constrained equations as presented in Section 4.3.1, the work [72] presents the method of interior points, work [73] describes the sphere decoding algorithm, and the work [74] presents a multi-parametric programming technique for solving it.

Chapter 5

Modulated FCS-MPC for DFIG-DPC

Alfeu J. Sguarezi Filho[a] and Rogério V. Jacomini[b]

[a]*Federal University of ABC, Santo André, SP, Brazil,* [b]*Federal Institute of São Paulo, Hortolândia, SP, Brazil*

5.1 Representation of DFIG using DPC

Section 3.1.4.1 introduces the fundamental concepts of DPC using SVOC, and this section is presented in the synchronous reference frame. In this way, stator and rotor flux elements are employed and no element is neglected initially.

The apparent complex power ($S_s = P + jQ$) in Eq. (2.54) representing voltages and currents is

$$\vec{S}_s = \overbrace{\frac{3}{2}\mathcal{R}e(\vec{v}_{s,dq} \cdot \vec{i}^{\,*}_{s,dq})}^{P} + j\,\underbrace{\frac{3}{2}\mathcal{I}m(\vec{v}_{s,dq} \cdot \vec{i}^{\,*}_{s,dq})}_{Q}. \tag{5.1}$$

The stator and rotor flux vectors have a relation that can be represented by manipulating Eqs. (2.50)–(2.51), and the stator current vector can be represented as

$$\vec{i}_{s,dq} = \frac{\vec{\lambda}_{s,dq}}{\sigma L_s} - \frac{k_r}{\sigma L_s}\vec{\lambda}_{r,dq}, \tag{5.2}$$

where $\sigma = (L_s L_r - L_m^2)/L_s L_r$ is the total leakage factor and $k_r = L_m/L_r$.

The reactive and active power can be represented by means of Eq. (5.2) substituted into Eq. (5.1):

$$P = -k_\sigma v_{s,q}\left(\lambda_{r,q} - \frac{\lambda_{s,q}}{k_r}\right), \tag{5.3}$$

$$Q = -k_\sigma v_{s,q}\left(\lambda_{r,d} - \frac{\lambda_{s,d}}{k_r}\right), \tag{5.4}$$

where $k_\sigma = 1.5k_r/\sigma L_s$ and $k_s = L_m/L_s$.

5.1.1 Rotor voltage representation

The rotor current vector can be expressed using the flux linkage vectors in Eqs. (2.50) and (2.51) as follows:

https://doi.org/10.1016/B978-0-32-390964-8.00014-2

$$\vec{i}_{r,dq} = \frac{\vec{\lambda}_{r,dq}}{\sigma L_r} - \frac{k_s}{\sigma L_r}\vec{\lambda}_{s,dq}. \tag{5.5}$$

Substituting Eq. (5.5) into Eq. (2.48), the elements of the rotor voltage vector can be expressed by

$$v_{r,d} = \frac{d\lambda_{r,d}}{dt} - \omega_{sl}\lambda_{r,q} + \frac{R_r}{\sigma L_r}\lambda_{r,d} - \frac{R_r k_s}{\sigma L_r}\lambda_{s,d}, \tag{5.6}$$

$$v_{r,q} = \frac{d\lambda_{r,q}}{dt} + \omega_{sl}\lambda_{r,d} + \frac{r_r}{\sigma L_r}\lambda_{r,q} - \frac{R_r k_s}{\sigma L_r}\lambda_{s,q}. \tag{5.7}$$

To compute the required elements of the rotor voltage vector, the rotor can be represented by using the stator reactive and active power. In this way, rearranging Eqs. (5.3) and (5.4), it is possible to obtain the expressions as follows:

$$\lambda_{r,q} = \frac{\lambda_{s,q}}{k_r} - \frac{P}{k_\sigma v_{s,q}}, \tag{5.8}$$

$$\lambda_{r,d} = \frac{\lambda_{s,d}}{k_r} - \frac{Q}{k_\sigma v_{s,q}}. \tag{5.9}$$

And replacing into Eqs. (5.6) and (5.7), considering $\lambda_{s,q} = 0$ and $\lambda_{s,d}$ constant, i.e., $\frac{d(\lambda_{s,d})}{dt} = 0$ [75], the following expressions are obtained:

$$v_{r,d} = -\frac{1}{k_\sigma}\frac{d\,(Q/v_{s,q})}{dt} + \frac{\omega_{sl}}{k_\sigma v_{s,q}}P - \frac{k_1}{v_{s,q}}Q + k_2\lambda_{s,d}, \tag{5.10}$$

$$v_{r,q} = -\frac{1}{k_\sigma}\frac{d\,(P/v_{s,q})}{dt} - \frac{k_1}{v_{s,q}}P - \frac{\omega_2}{k_\sigma v_{s,q}}Q + \frac{\omega_2}{k_r}\lambda_{s,d}, \tag{5.11}$$

where $k_1 = R_r/(\sigma L_r k_\sigma)$ and $k_2 = R_r(1 - k_s k_r)/(\sigma L_r k_r)$.

When a disturbance occurs in the grid, such as a voltage sag, the instantaneous value of the stator flux elements in the dq frame vary and, therefore, the stator voltages can be expressed as follows:

$$d\,(Q/v_{s,q})/dt \neq (1/v_{s,q})d\,Q/dt, \tag{5.12}$$

$$d\,(P/v_{s,q})/dt \neq (1/v_{s,q})d\,P/dt. \tag{5.13}$$

These equations in the expanded form can be represented as:

$$\frac{d\,(Q/v_{s,q})}{dt} = \frac{1}{v_{s,q}}\frac{d\,Q}{dt} - \frac{Q}{v_{s,q}^2}\frac{d\,v_{s,q}}{dt}, \tag{5.14}$$

$$\frac{d\,(P/v_{s,q})}{dt} = \frac{1}{v_{s,q}}\frac{d\,P}{dt} - \frac{P}{v_{s,q}^2}\frac{d\,v_{s,q}}{dt}. \tag{5.15}$$

The elements of the rotor voltage vector in dq employing Eqs. (5.14) and (5.15) in Eqs. (5.10) and (5.11) can be represented as follows:

$$v_{r,d} = -\frac{1}{k_\sigma v_{s,q}}\frac{dQ}{dt} + \frac{\omega_{sl}}{k_\sigma v_{s,q}}P - \frac{k_1}{v_{s,q}}Q + k_2\lambda_{s,d} + \frac{Q}{k_\sigma v_{s,q}^2}\frac{dv_{s,q}}{dt}, \tag{5.16}$$

$$v_{r,q} = -\frac{1}{k_\sigma v_{s,q}}\frac{dP}{dt} - \frac{k_1}{v_{s,q}}P_s - \frac{\omega_{sl}}{k_\sigma v_{s,q}}Q + \frac{\omega_{sl}}{k_r}\lambda_{s,d} - \frac{P}{k_\sigma v_{s,q}^2}\frac{dv_{s,q}}{dt}. \tag{5.17}$$

5.2 DPC for DFIG using the modulated FCS-MPC

DPC using the modulated FCS-MPC for DFIG strategy was inspired by the work presented [75] by the same authors of this chapter, in which the modulated FCS-MPC solution was not presented. So, the main objective is to compute the rotor voltage vector components ($v_{r,d}$ and $v_{r,q}$) using SVOC. This goal can be achieved expressing the rotor equations in discrete representation using the Euler method $\left(\frac{dx(t)}{dt} \cong \frac{x(k+1)-x(k)}{T}\right)$ in Eqs. (5.16) and (5.17). Hence, the discrete rotor equations for a sampling period T are given by

$$\begin{bmatrix} v_{r,d}(k+1) \\ v_{r,q}(k+1) \end{bmatrix} = -\frac{1}{Tk_\sigma}\cdot\begin{bmatrix} \frac{1}{v_{s,q}(k)} & 0 \\ 0 & \frac{1}{v_{s,q}(k)} \end{bmatrix}\cdot\begin{bmatrix} \Delta Q(k) \\ \Delta P(k) \end{bmatrix}$$
$$+\begin{bmatrix} -\frac{k_1}{v_{s,q}} & \frac{\omega_{sl}(k)}{k_\sigma v_{s,q}} \\ -\frac{\omega_{sl}(k)}{k_\sigma v_{s,q}} & -\frac{k_1}{v_{s,q}} \end{bmatrix}\cdot\begin{bmatrix} Q(k) \\ P(k) \end{bmatrix} + \begin{bmatrix} 1 & 0 \\ 0 & 1 \end{bmatrix}\cdot\begin{bmatrix} k_2 \\ \frac{\omega_{sl}(k)}{k_r} \end{bmatrix}\cdot\lambda_{s,d}(k)$$
$$+\frac{1}{Tk_r}\begin{bmatrix} 1 & 0 \\ 0 & 1 \end{bmatrix}\cdot\begin{bmatrix} \frac{Q(k)}{v_{s,q}^2(k)} \\ \frac{P(k)}{v_{s,q}^2(k)} \end{bmatrix}\cdot\Delta v_{sq}, \tag{5.18}$$

where the variation of the quadrature element of the stator voltage at the sampling instant $k-th$ is represented by $\Delta v_{s,q}(k) = v_{s,q}(k+1) - v_{s,q}(k)$; the errors of the reactive and active power are expressed as $\Delta Q(k) = Q_{ref}(k+1) - Q(k)$ and $\Delta P(k) = P_{ref}(k+1) - P(k)$, respectively.

DPC using modulated FCS-MPC for DFIG predicts the rotor voltage elements using Eq. (5.18) and the references of the apparent power ($P_{ref}(k+1)$ and reactive $Q_{ref}(k+1)$). In this way, the predicted rotor voltage elements in the $\alpha\beta r$ frame can be represented as

$$\vec{v}_{r,\alpha\beta r}(k+1) = \vec{v}_{r,dq}(k+1)\, e^{j\theta_{sl}}. \tag{5.19}$$

The rotor voltage vector presented in Eq. (5.19) is employed to determine the switching states of optimal rotor voltage vectors inspired in the works presented in [76,77]. The possible values of $v_{r,\alpha r}(k)$ and $v_{r,\beta r}(k)$ can be seen in Table 5.1

and the information of this table is in accordance with the states of the converter, as presented in Table 2.1. Initially, the first voltage rotor vector is selected using the minimized cost function which can be expressed as

$$J_{dpc} = |v_{r,\beta r}(h) - v_{r,\beta r}(k+1)| + |v_{r,\alpha r}(h) - v_{r,\alpha r}(k+1)|. \tag{5.20}$$

TABLE 5.1 Switching states and voltage vectors.

S_a	S_b	S_c	Rotor voltage vectors ($\vec{v}_{r,\alpha\beta r}(h)$)
1	0	0	$\vec{v}_r(1) = \frac{2}{3}V_{dc}$
1	1	0	$\vec{v}_r(2) = \frac{1}{3}V_{dc} + j\frac{\sqrt{3}}{3}V_{dc}$
0	1	0	$\vec{v}_r(3) = -\frac{1}{3}V_{dc} + j\frac{\sqrt{3}}{3}V_{dc}$
0	1	1	$\vec{v}_r(4) = -\frac{2}{3}V_{dc}$
0	0	1	$\vec{v}_r(5) = -\frac{1}{3}V_{dc} - j\frac{\sqrt{3}}{3}V_{dc}$
1	0	1	$\vec{v}_r(6) = \frac{1}{3}V_{dc} - j\frac{\sqrt{3}}{3}V_{dc}$

The angle of the rotor voltage vector in the $\alpha\beta r$ frame is employed to determine the sector using Table 5.2, and part from this sector and the vector determined in Eq. (5.20) are obtained from two adjacent vectors, as shown in Fig. 5.1.

TABLE 5.2 Sector of reference rotor voltage vector.

Angle range	Sector
$0^\circ \leq \theta_{sl}(k) < 60^\circ$	I
$60^\circ \leq \theta_{sl}(k) < 120^\circ$	II
$120^\circ \leq \theta_{sl}(k) < 180^\circ$	III
$180^\circ \leq \theta_{sl}(k) < 240^\circ$	IV
$240^\circ \leq \theta_{sl}(k) < 300^\circ$	V
$300^\circ \leq \theta_{sl}(k) < 360^\circ$	VI

In this manner, the smallest value of the minimized cost function J_{dpc} is chosen between the two adjacent vectors which permits determining the values of the duty cycles on pulse width switching. Figs. 5.2 and 5.3 depict the block diagram of the DPC for DFIG using modulated FCS-MPC. The basic code of the modulated FCS-MPC strategy is presented in Algorithm 5.1 and the computation of the duty cycles are the goal of the next section.

5.2.1 Computation of the duty cycle using modulated FCS-MPC

In this section, for instance, it is assumed that the command vector $\vec{v}_{r,\alpha\beta r}$ is in sector I and the switching sequence implemented is $\vec{v}_r(0)$, $\vec{v}_r(1)$, $\vec{v}_r(2)$,

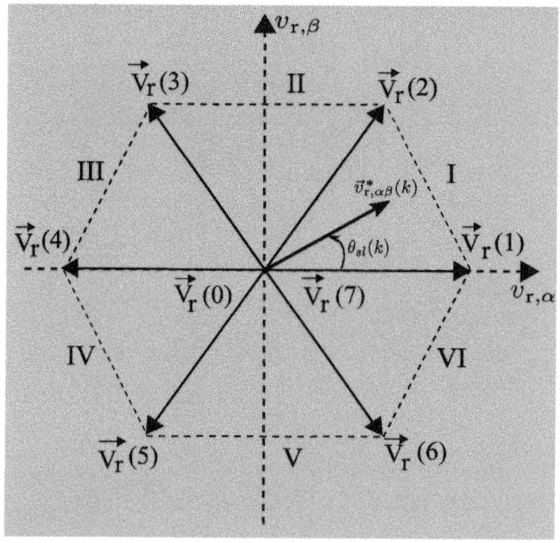

FIGURE 5.1 Graphical representation of converter switching conditions and the rotor voltage vector in the rotor reference.

$\vec{v}_r(7)$, $\vec{v}_r(2)$, $\vec{v}_r(1)$, and $\vec{v}_r(0)$. In the mentioned sequence, the period of application of each switching vector ($\vec{v}_r(1)$, $\vec{v}_r(2)$, $\vec{v}_r(0)/\vec{v}_r(7)$) in a period T is determined. For the average voltage made by the converter in time T to be the same to $\vec{v}_{r,\alpha\beta r}$ predicted using the modulated FCS-MPC, the following equation must be satisfied:

$$\vec{v}_{r,\alpha\beta r} = \frac{1}{T}\left(\int_{t_1}^{0} \vec{v}_r(0)dt + \int_{t_1}^{t_2} \vec{v}_r(1)dt + \int_{t_2}^{T} \vec{v}_r(2)dt\right). \tag{5.21}$$

As $\vec{v}_r(0)/\vec{v}_r(7)$ are null vectors, Eq. (5.21) can be rewritten as

$$\vec{v}_{r,\alpha\beta r} = \frac{1}{T}[\vec{v}_r(1)(t_2 - t_1) + \vec{v}_r(2)(T - t_2)]. \tag{5.22}$$

Defining the application of the time interval of the vector $\vec{v}_r(1)$ as Δt_1, the application of the time interval of the vector $\vec{v}_r(2)$ as Δt_2, and the application of the time interval of the vector $\vec{v}_r(0)$ as Δt_0, it can be represented as

$$\vec{v}_{r,\alpha\beta r} = \frac{1}{T}[\vec{v}_r(2)\Delta t_2 + \vec{v}_r(1)\Delta t_1]. \tag{5.23}$$

In this way, by defining the duty cycle for each of the mentioned vectors, that is, the reason for the time each of them must be applied over the period T, it is possible to obtain:

$$\vec{v}_{r,\alpha\beta r} = D_1\vec{v}_r(1) + D_2\vec{v}_r(2), \tag{5.24}$$

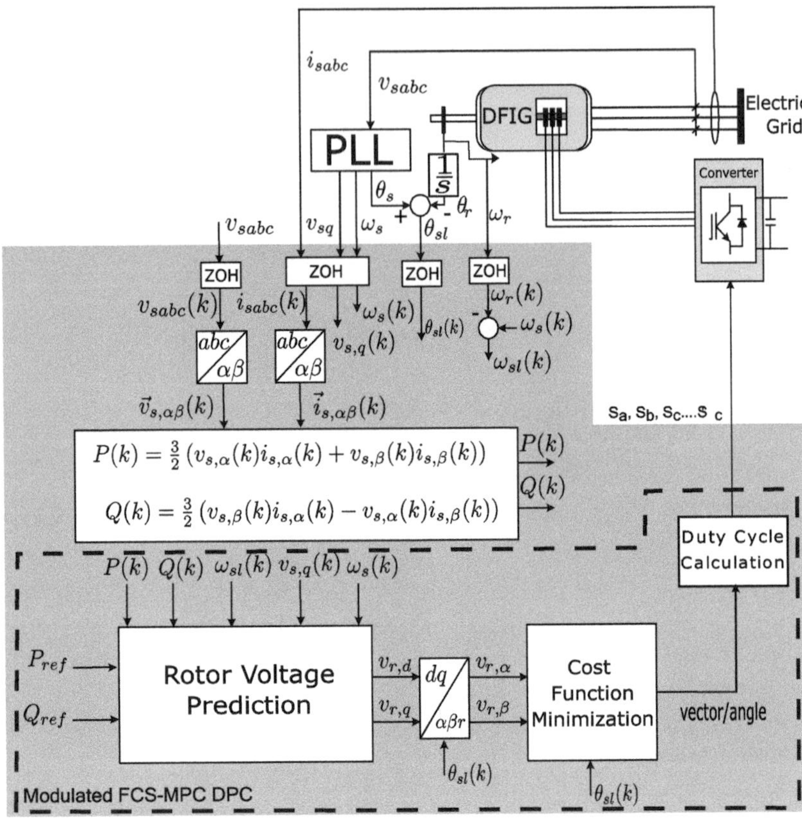

FIGURE 5.2 DPC for DFIG using modulated FCS-MPC block scheme of the implemented control algorithm.

where $D_1 = \Delta t_1/T$ and $D_2 = \Delta t_2/T$. As $T = \Delta t_1 + \Delta t_2 + \Delta t_0$, one can also state that

$$D_1 + D_2 + D_0 = 1. \tag{5.25}$$

Eq. (5.24) is expressed in matrix form as follows:

$$\vec{v}_{r,\alpha\beta r} = \begin{bmatrix} \vec{v}_r(1) \\ \vec{v}_r(2) \end{bmatrix} \begin{bmatrix} D_1 \\ D_2 \end{bmatrix}. \tag{5.26}$$

By writing the vectors involved in the $\alpha\beta r$ coordinate frame, one gets

$$\begin{bmatrix} v_{r,\alpha r} \\ v_{r,\beta r} \end{bmatrix} = \begin{bmatrix} v_1^{\alpha} & v_2^{\alpha} \\ v_1^{\beta} & v_2^{\beta} \end{bmatrix} \begin{bmatrix} D_1 \\ D_2 \end{bmatrix}. \tag{5.27}$$

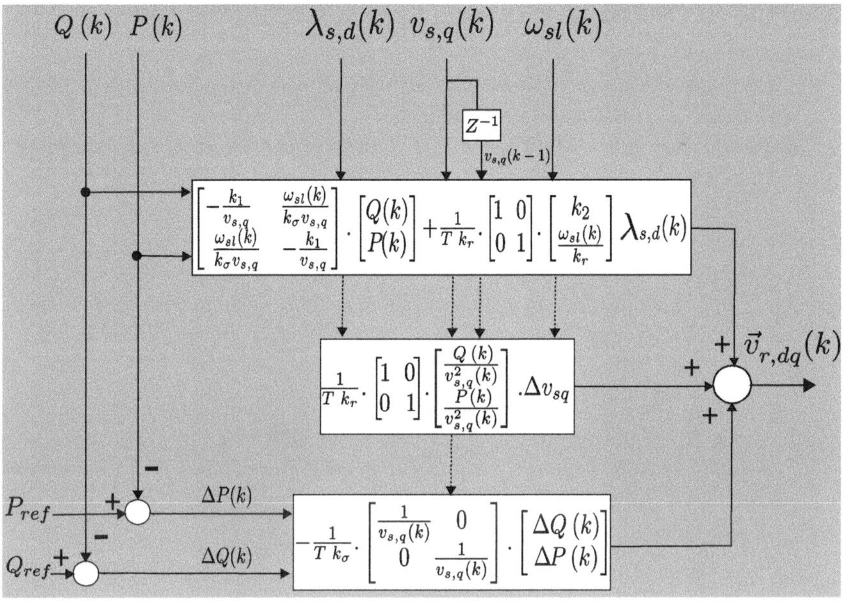

FIGURE 5.3 DPC for DFIG using modulated FCS-MPC block scheme of the rotor voltage prediction.

Finally, one can add one more row and one more column to the 2×2 matrix of Eq. (5.25) with the equality presented in Eq. (5.27), to obtain a single matrix equation involving the duty cycle of all vectors used in the implementation. Hence, the new representation is reached:

$$\begin{bmatrix} v_{r,\alpha r} \\ v_{r,\beta r} \\ 1 \end{bmatrix} = \begin{bmatrix} v_1^{\alpha} & v_2^{\alpha} & v_0^{\alpha} \\ v_1^{\beta} & v_2^{\beta} & v_0^{\beta} \\ 1 & 1 & 1 \end{bmatrix} \begin{bmatrix} D_1 \\ D_2 \\ D_0 \end{bmatrix}. \tag{5.28}$$

Generalizing Eq. (5.28) for all six sectors gives

$$\begin{aligned} \begin{bmatrix} v_{r,\alpha r} \\ v_{r,\beta r} \\ 1 \end{bmatrix} &= \begin{bmatrix} v_{i-mfcs}^{\alpha} & v_{j-mfcs}^{\alpha} & v_0^{\alpha} \\ v_{i-mfcs}^{\beta} & v_{j-mfcs}^{\beta} & v_0^{\beta} \\ 1 & 1 & 1 \end{bmatrix} \begin{bmatrix} D_{i-mfcs} \\ D_{j-mfcs} \\ D_0 \end{bmatrix} \\ &= M_{k-mfcs}^{-1} \begin{bmatrix} D_{i-mfcs} \\ D_{j-mfcs} \\ D_0 \end{bmatrix} \end{aligned} \tag{5.29}$$

where $i - mfcs$ and $j - mfcs$ represent the indexes of the adjacent nonnull vectors for sector $k - mfcs$, and M_{k-mfcs} is called the sector $k - mfcs$ decomposition matrix.

Algorithm 5.1 DPC for DFIG using Modulated MPC-FCS algorithm

while MPC-FCS active **do**
  The system variables such as current, stator flux, reactive and active power, stator voltage, rotor speed are measured
  $Sector$ = function of θ_{sl} (Table 5.2) % determines the sector in function of rotor position
  $i - mfcs = Sector$ % Obtains the adjacent vectors %($i - mfcs$ and $j - mfcs$) in function of sector
  $j - mfcs = Sector + 1$
  $J_{dpc-op} = \infty$
  for
  h = i : j **do**
    $\vec{v_r}(h)$ = function of V_{dc} % Possible states of the converter
    $\vec{v}_{r,\alpha\beta r}(k) = \vec{v}_{r,dq}(k)e^{j(\theta_s-\theta_r)}$ % System prediction
    $J_{dpc}(h) = |v_{r,\alpha r}(h) - v_{r,\alpha r}(k+1)| + |v_{r,\beta r}(h) - v_{r,\beta r}(k+1)|$ % Cost function calculation
    if $J_{dpc}(h) < J_{dpc-op}$ **then**
      $J_{dpc-op} = J_{dpc}(h)$
      $J_{dpc-op} = h$ % Store the state with the least cost
    end if
    if $h = 2$ **then**
      Apply $\vec{v_r}(J_{dpc-op})$
      Start again
      Else go to **for** loop again
    end if
  end for
  D0 = (Table 5.3) % calculation of the duty cycle
  D1 = (Table 5.3) % calculation of the duty cycle
  D2 = (Table 5.3) % calculation of the duty cycle
end while

Finally, the duty cycle of the adjacent vectors for each sector can be calculated from the M_{k-mfcs} decomposition matrix, as shown in

$$\begin{bmatrix} D_{i-mfcs} \\ D_{j-mfcs} \\ D_0 \end{bmatrix} = M_{k-mfcs} \begin{bmatrix} v_{r,\alpha r} \\ v_{r,\beta r} \\ 1 \end{bmatrix}. \tag{5.30}$$

The calculation of the duty cycle of the vectors in the same sector to each of the sectors was accomplished by using Table 5.3, where $X_{mfcs} = \frac{3}{2} v_{r,\alpha r} - \frac{\sqrt{3}}{2} v_{r,\beta r}$, $Y_{mfcs} = \frac{3}{2} v_{r,\alpha r} + \frac{\sqrt{3}}{2} v_{r,\beta r}$, and $Z_{mfcs} = \sqrt{3}\, v_{r,\beta r}$.

TABLE 5.3 Calculation of the duty cycle of the adjacent vectors to each sector $k - mfcs$ in the $\alpha\beta r$ frame.

Sector	D_{i-mfcs}	D_{j-mfcs}	D_0
I	$D_1 = X_{mfcs}$	$D_2 = Z_{mfcs}$	$-Y_{mfcs} + 1$
II	$D_2 = Y_{mfcs}$	$D_3 = -X_{mfcs}$	$-Z_{mfcs} + 1$
III	$D_3 = Z_{mfcs}$	$D_4 = -Y_{mfcs}$	$X_{mfcs} + 1$
IV	$D_4 = -X_{mfcs}$	$D_5 = -Z_{mfcs}$	$Y_{mfcs} + 1$
V	$D_5 = -Y_{mfcs}$	$D_6 = X_{mfcs}$	$Z_{mfcs} + 1$
VI	$D_6 = -Z_{mfcs}$	$D_1 = Y_{mfcs}$	$-X_{mfcs} + 1$

5.3 Experimental results

This section introduces the experimental results obtained in the setup using the data presented in Table A.1, and the rotor speed is maintained by the DC motor at 16750 rpm.

The behavior of the reactive and active power during several step tests is depicted in Fig. 5.4, and in this test the stator active power (Fig. 5.4(a)) changes from -2 to -1 kW and changes again to 1.5 kW, and the stator reactive power (Fig. 5.4(b)) changes from -0 to -600 var and changes again to 1 kvar. It can be noticed that the DPC for DFIG using the modulated FCS-MPC calculates the rotor voltage vector necessary for the stator power to reach its references. The settling time is about 1 ms and it can be seen in Fig. 5.5. The oscillations occur due to the switching of the inverter.

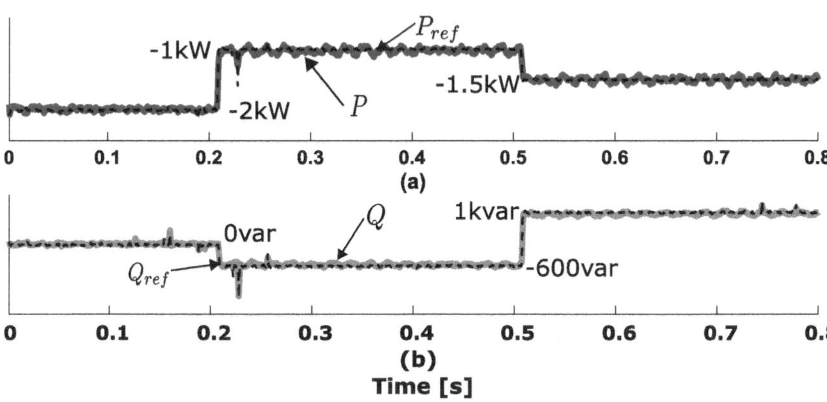

FIGURE 5.4 Modulated FCS-MPC DPC. Step tests for apparent power: (a) active power and (b) reactive power.

The performance of the stator current and voltage during the several step tests for the power references is depicted in Figs. 5.6 and 5.7, respectively. In both cases, (a) is related to the stator apparent power and (b) is related to the

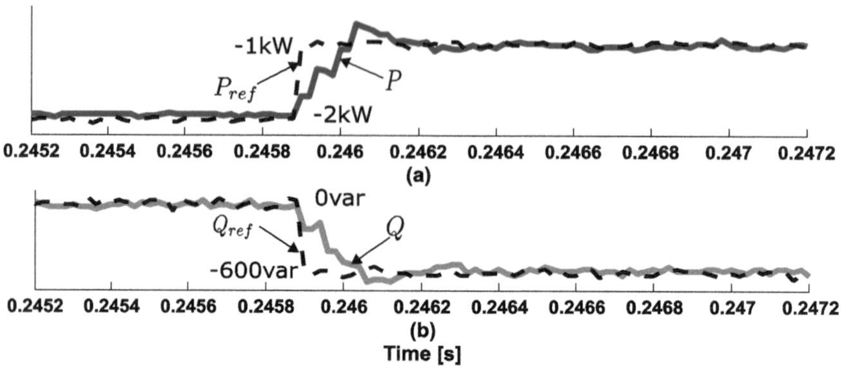

FIGURE 5.5 Modulated FCS-MPC DPC. Detail of the step test for apparent power: (a) active power and (b) reactive power.

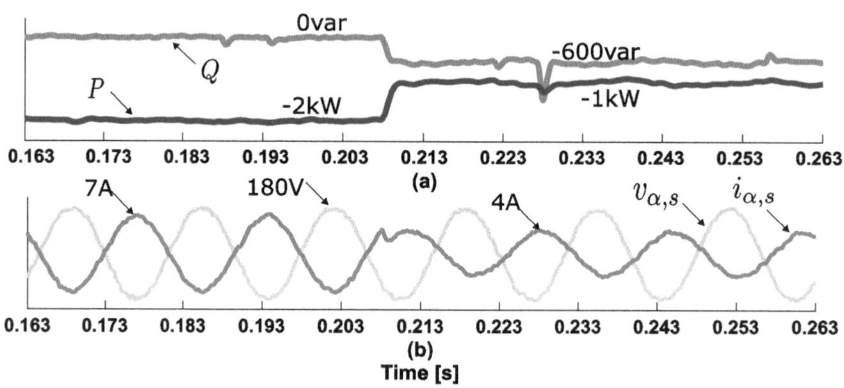

FIGURE 5.6 Modulated FCS-MPC DPC. Detail of the step test for apparent power: (a) reactive power and active power, (b) stator current and voltage.

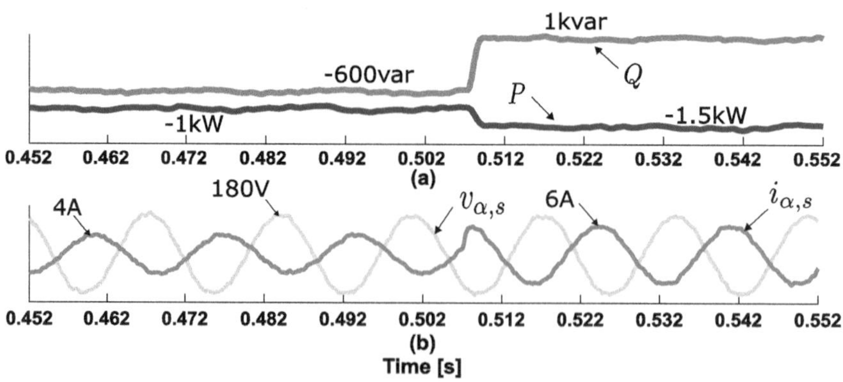

FIGURE 5.7 Modulated FCS-MPC DPC. Detail of the step test for apparent power: (a) reactive power and active power, (b) stator current and voltage.

stator voltage and current. It can be seen that the stator current increases its amplitude when the active power changes its value, and the stator current changes its angle when the reactive power changes its value due to the fact the stator current vector in the dq frame is related to the active and reactive stator power in accordance with the model representation. Finally, the THD of the rotor voltage during the test is depicted in Fig. 5.8 and it endorses the fixed frequency operation.

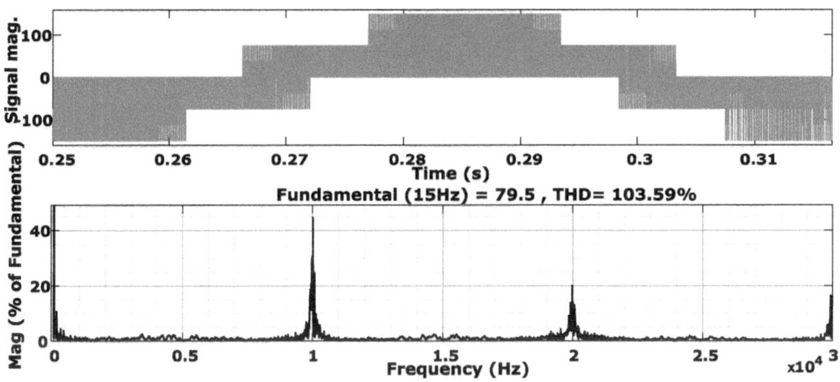

FIGURE 5.8 Modulated FCS-MPC DPC. THD of the rotor voltage during the test.

In the last test, DFIG changes its speed from 1710 to 1980 rpm, the stator reactive power changes from 0 to −600 var and again to 1 kvar, and the stator active power changes from −1 to −1.5 kW and again to −2 kW. The actions of the speed of the rotor, the stator apparent power elements and the rotor current in the $\alpha\beta r$ frame are depicted in Figs. 5.9–5.11, respectively. It can be noticed that the DPC for DFIG by using modulated FCS-MPC reaches the references during the mentioned test.

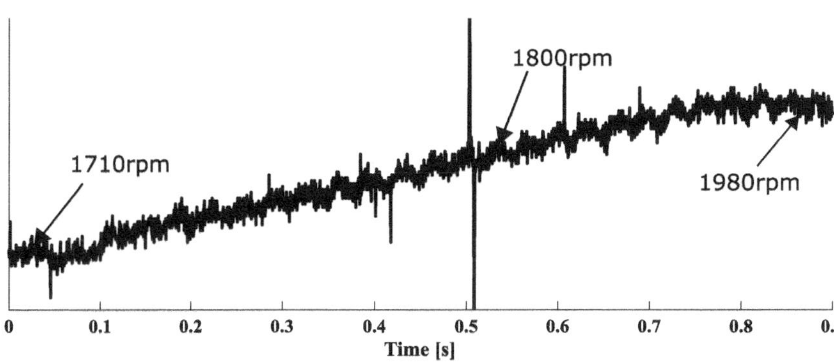

FIGURE 5.9 Modulated FCS-MPC DPC. Rotor speed.

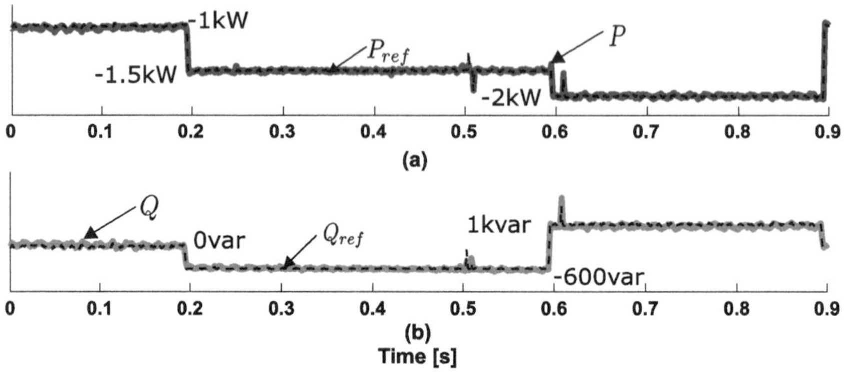

FIGURE 5.10 Modulated FCS-MPC DPC. Step test for apparent power: (a) stator active power and (b) stator reactive power.

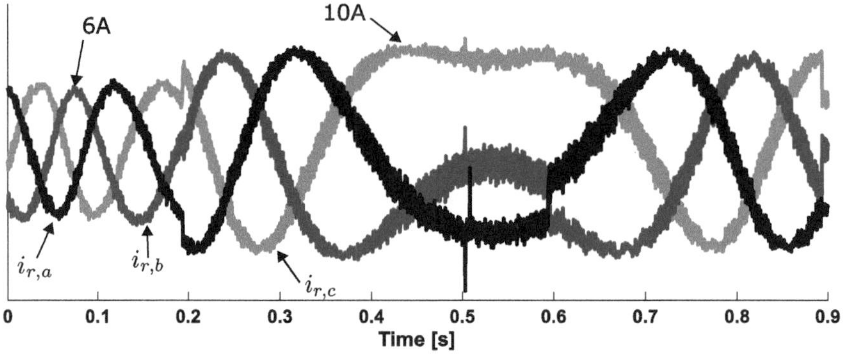

FIGURE 5.11 Modulated FCS-MPC DPC. Rotor current during several-speed operation.

5.4 Summary

This chapter describes the DPC for DFIG using the modulated FCS-MPC technique in normal operation conditions. The modulated FCS-MPC employs the discrete representation of the DFIG in the dq frame. In both cases, the modulated FCS-MPC could reach the references of power in accordance with the presented experimental results. Additionally, this technique operates using constant switching frequency and it may be of interest to DPC for DFIG.

Chapter 6

A wireless coded modulated FCS-MPC DPC for renewable energy sources in smart grid environment

Alfeu J. Sguarezi Filho, Angelo S. Lunardi, Carlos E. Capovilla, and Ivan R.S. Casella
Federal University of ABC, Santo André, SP, Brazil

6.1 Overview

The concept of smart grids (SGs) is the evolution of the current power model, which is composed of an infrastructure that brings benefits for all systems involved, such as distribution, generation of energy, consumption, and transmission. The modern information and communication technologies permit the control and monitoring of the energy grid. In this way, the reliability, efficiency, and security are increased [78,79].

The definition for SGs is a self-sufficient integrated system with an electrical grid; it can be found in any type or any scale, using different types of energy source generators. It can also be defined as an intelligent digitized system to provide optimized energy from source to consumer [80].

The SGs can be divided into three infrastructures. The first is the smart infrastructure that has other three subdivisions, the energy infrastructure which is responsible for generation, transmission, and consumption of energy. The information infrastructure is another subdivision of smart infrastructure, which supports information generation, modeling, integration, analysis, and optimization. The last subdivision is the communication infrastructure, which is responsible for communication and transmission of information throughout the system [81].

The second infrastructure for SGs is the smart management system, all the advanced management and control of services and features are under the responsibility of this system. The management of technologies and services is developed through this application in a more intelligent and efficient way [82].

Finally, the smart protection system is the last infrastructure, which provides an analysis of grid reliability, failure protection, security services, and privacy

https://doi.org/10.1016/B978-0-32-390964-8.00015-4

protection [78]. Fig. 6.1 presents a diagram for a conventional smart grid infrastructure.

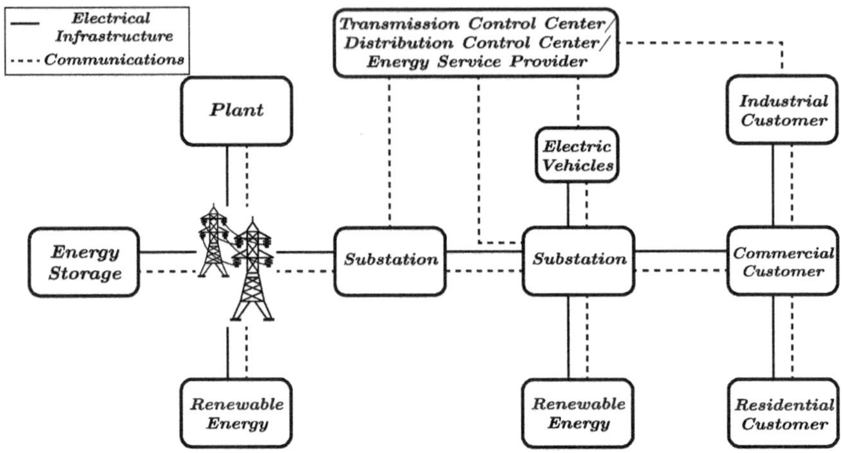

FIGURE 6.1 Smart grid infrastructure.

The success of an SG system is a result of using the concept of multiple renewable energy sources (RES) for energy conversion technology. There are several types of RES, such as storage, wind, and solar, connected to the electrical grid. And this type of operation is allowed due to the power electronics converters [83]. Therefore, it permits increasing the reliability, efficiency, security, and control flexibility of the power system [84,85]. The capability of the three-phase inverter to inject power is the solution for closed-loop control [86,87].

The wireless communication systems are statistically susceptible to errors even if they are a useful key to the SG [88,89]. The forward error correction (FEC) techniques are commonly used due the fact that they permit minimizing the error probability. This type of solution is inserted in most modern wireless digital systems and it permits ensuring the reception of the transmitted information [90,91].

Among them, low density parity check (LDPC) coding presents an excellent performance and shows an optimized compromise between decoding complexity and performance [91]. The LDPC coding is one of the most efficient coding schemes used in practice for high data rate transmissions in fading channels with low signal-to-noise ratio (E_b/N_0) [91]. Specifically, the eIRA coding, employed in this section, is a special class of LDPC that integrates the DVB-S2 and WiMAX standards, and presents a relative low encoding/decoding complexity (e.g., eIRA codes offer lower decoding complexity than TC (turbo coding)).

Such modern wireless systems are also based on orthogonal frequency division multiplexing (OFDM) transmission technique. Due to its resistance to intersymbol interference (ISI) and efficient implementation, OFDM is the technology for modern and future communication systems [92,93].

The work presented in this chapter is based in the work [94]. In this case, the same wireless OFDM-LDPC system was employed to transmit the power references to the converter from the SG operator. However, the modulated FCS-MPC DPC for the three-phase power CCG and the cost function using grid current were employed in this configuration, new analysis and experimental results are inserted.

6.2 Three-phase power CCG using direct power predictive control

In this section, control of a converter connected to the electrical grid applying the direct power control using modulated FCS-MPC will be presented. The mathematical expression of the three-phase power CCG was presented in Section 2.3.2 and by Eqs. (2.65) and (2.66) that present the predictive values for grid current:

$$\vec{i}_{g,\alpha\beta}(k+1) = \left(1 - \frac{TR_g}{L_g}\right)\vec{i}_{g,\alpha\beta}(k) + \frac{T}{L_g}(\vec{v}_{invg,\alpha\beta} - \vec{v}_{g,\alpha\beta}(k)), \quad (6.1)$$

where each element of the vector can be expressed as

$$i_{g,\alpha}(k+1) = \left(1 - \frac{TR_g}{L_g}\right) i_{g,\alpha}(k) + \frac{T}{L_g}(v_{invg,\alpha} - v_{g,\alpha}(k)) \quad (6.2)$$

and

$$i_{g,\beta}(k+1) = \left(1 - \frac{TR_g}{L_g}\right) i_{g,\beta}(k) + \frac{T}{L_g}(v_{invg,\beta} - v_{g,\beta}(k)). \quad (6.3)$$

The prediction of the grid current vector will be carried out by means of Eqs. (6.1)–(6.3) can be employed to predict the power behavior. In this case, this strategy employs eight possible states for an inverter as presented in Table 6.1 in accordance with Fig. 6.2.

TABLE 6.1 Switching states and voltage vectors.

S_a	S_b	S_c	Voltage vector $\vec{v}_g$
0	0	0	$\vec{v}_g(0) = 0$
1	0	0	$\vec{v}_g(1) = \frac{2}{3}V_{dc}$
1	1	0	$\vec{v}_g(2) = \frac{1}{3}V_{dc} + j\frac{\sqrt{3}}{3}V_{dc}$
0	1	0	$\vec{v}_g(3) = -\frac{1}{3}V_{dc} + j\frac{\sqrt{3}}{3}V_{dc}$
0	1	1	$\vec{v}_g(4) = -\frac{2}{3}V_{dc}$
0	0	1	$\vec{v}_g(5) = -\frac{1}{3}V_{dc} - j\frac{\sqrt{3}}{3}V_{dc}$
1	0	1	$\vec{v}_g(6) = \frac{1}{3}V_{dc} - j\frac{\sqrt{3}}{3}V_{dc}$
1	1	1	$\vec{v}_g(7) = 0$

The current references can be obtained from the active and reactive reference powers presented in Eq. (6.4), after algebraic manipulation and using the following equations:

$$\begin{bmatrix} i_{g,\alpha\ ref} \\ i_{g,\beta\ ref} \end{bmatrix} = \frac{2}{3(v_{g,\alpha}^2(k) + v_{g,\beta}^2(k))} \begin{bmatrix} v_{g,\alpha}(k) & v_{g,\beta}(k) \\ v_{g,\beta}(k) & -v_{g,\alpha}(k) \end{bmatrix} \begin{bmatrix} P_{g\ ref} \\ Q_{g\ ref} \end{bmatrix}. \quad (6.4)$$

Hence, it is possible to control the reactive and active power by means the grid current vector control.

The first cost function which should be solved for each pair of corresponding vector in Fig. 6.2 and Table 6.1 is expressed in

$$J_{dpcccg} = |i_{g,\alpha\ ref} - i_{g,\alpha}(k+1)| + |i_{g,\beta\ ref} - i_{g,\beta}(k+1)|. \quad (6.5)$$

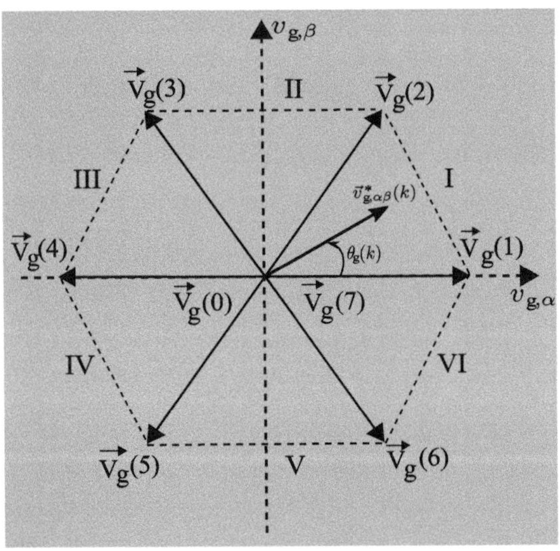

FIGURE 6.2 VSI sextant.

Fig. 6.2 refers to the voltage vectors available for the VSI in the $\alpha\beta$ frame, in which one is used in space vector modulation (SVM). Fig. 6.2 is divided into six sectors with their related two adjacent vectors. The first sector, for example, is between vector $\vec{v}_g(1)$ and vector $\vec{v}_g(2)$ presented in Table 6.1 and its respectively a voltage generated vector. Based on the classical predictive control method, a method is presented in Fig. 6.3 that uses the same prediction of the load current indicated in Eqs. (6.2) and (6.3). Adding the main contribution, the proposed techniques use two active vectors in order to maintain a fixed switching frequency; to do this, it is necessary to evaluate the predictive cost functions separately.

First, the cost function J_{dpcccg} is found for each vector presented in VSI according to Table 6.1. As an example, for sector II, the cost function is evaluated for both vectors $\vec{v}_g(2)$ and $\vec{v}_g(3)$ based on Eqs. (6.2) and (6.3) that will change according to the grid voltage. Finally, the duty cycle for each vector is calculated by:

$$d_0 = \frac{K}{J_{0-dpcccg}}, \quad d_1 = \frac{K}{J_{1-dpcccg}}, \quad d_2 = \frac{K}{J_{2-dpcccg}}, \tag{6.6}$$

$$d_0 + d_1 + d_2 = 1, \tag{6.7}$$

where d_0 is the duty cycle of a zero vector ($\vec{v}_g(0)$ or $\vec{v}_g(7)$) that should be evaluated just once. It is possible to obtain the equation for K solving the system of Eq. (6.6), and the solution is represented as follows:

$$\begin{aligned} d_0 &= \frac{J_{1-dpcccg} J_{2-dpcccg}}{J_{0-dpcccg} J_{1-dpcccg} + J_{0-dpcccg} J_{2-dpcccg} + J_{1-dpcccg} J_{2-dpcccg}}, \\ d_1 &= \frac{J_{0-dpcccg} J_{2-dpcccg}}{J_{0-dpcccg} J_{1-dpcccg} + J_{0-dpcccg} J_{2-dpcccg} + J_{1-dpcccg} J_{2-dpcccg}}, \\ d_2 &= \frac{J_{0-dpcccg} J_{1-dpcccg}}{J_{0-dpcccg} J_{1-dpcccg} + J_{0-dpcccg} J_{2-dpcccg} + J_{1-dpcccg} J_{2-dpcccg}}. \end{aligned} \tag{6.8}$$

From the predicted actions employing Eqs. (6.2) and (6.3), the cost function is calculated in order to select the voltage vectors of the inverter that allows the modulated FCS-MPC as presented in Section 4.2.3:

$$J_{dpcccgK}(k+1) = J_{1-dpcccg} d_1 + J_{2-dpcccg} d_2, \tag{6.9}$$

where $J_{1-dpcccg}$ and $J_{2-dpcccg}$ are the cost function with the least values and d_1 and d_2 are the duty cycles computed using Eq. (6.8). See Algorithm 6.1.

Algorithm 6.2 presents the implementation of the modulated FCS-MPC DPC for CCG, and the scheme of this predictive control system is shown in Fig. 6.3.

6.3 Representation of the wireless communication system

The scheme of the wireless communication system employed is depicted in Fig. 6.4.[1] This diagram links the reference grid reactive $\bar{Q}_{g,\alpha\beta ref}$ and active $\bar{P}_{g,\alpha\beta ref}$ power of the SG central control to the three-phase power CCG that processes the energy of the RES. The LDPC coding [95,96] and OFDM technique [90] are employed to transmit the references to ensure the higher reliance in frequency-selective fading channels.

LDPC coding strategies are generally based on a class of (binary) linear block codes with a sparse parity-check matrix H_{com} [96,97], whose parity-check matrix can be represented by

$$H_{com} = \begin{bmatrix} H_1 & H_2 \end{bmatrix}, \tag{6.10}$$

1 This figure is part of the work presented in [94].

Algorithm 6.1 SectorFunction algorithm

$d_0 = \dfrac{J_{1-dpcccg} J_{2-dpcccg}}{J_{0-dpcccg} J_{1-dpcccg} + J_{0-dpcccg} g_2 + J_{1-dpcccg} J_{2-dpcccg}}$

switch(Sector)

case:1

Obtain d_1 by Eq. (2.8) for $\vec{v}_g(1)$ and $\vec{v}_g(2)$ using respective $J_{1-dpcccg}$ and $J_{2-dpcccg}$

Obtain d_2 by Eq. (2.8) for $\vec{v}_g(1)$ and $\vec{v}_g(2)$ using respective $J_{1-dpcccg}$ and $J_{2-dpcccg}$

$J_{dpcccgK}[Sector] = J_{dpcccg}[1]d_1 + J_{dpcccg}[2]d_2$

end

case:2

Obtain d_1 by Eq. (2.8) for $\vec{v}_g(2)$ and $\vec{v}_g(3)$ using respective $J_{1-dpcccg}$ and $J_{2-dpcccg}$

Obtain d_2 by Eq. (2.8) for $\vec{v}_g(2)$ and $\vec{v}_g(3)$ using respective $J_{1-dpcccg}$ and $J_{2-dpcccg}$

$J_{dpcccgK}[Sector] = J_{dpcccg}[2]d_1 + J_{dpcccg}[3]d_2$

end

case:3

Obtain d_1 by Eq. (2.8) for $\vec{v}_g(3)$ and $\vec{v}_g(4)$ using respective $J_{1-dpcccg}$ and $J_{2-dpcccg}$

Obtain d_2 by Eq. (2.8) for $\vec{v}_g(3)$ and $\vec{v}_g(4)$ using respective $J_{1-dpcccg}$ and $J_{2-dpcccg}$

$J_{dpcccgK}[Sector] = J_{dpcccg}[3]d_1 + J_{dpcccg}[4]d_2$

end

case:4

Obtain d_1 by Eq. (2.8) for $\vec{v}_g(4)$ and $\vec{v}_g(5)$ using respective $J_{1-dpcccg}$ and $J_{2-dpcccg}$

Obtain d_2 by Eq. (2.8) for $\vec{v}_g(4)$ and $\vec{v}_g(5)$ using respective $J_{1-dpcccg}$ and $J_{2-dpcccg}$

$J_{dpcccgK}[Sector] = J_{dpcccg}[4]d_1 + J_{dpcccg}[5]d_2$

end

case:5

Obtain d_1 by Eq. (2.8) for $\vec{v}_g(5)$ and $\vec{v}_g(6)$ using respective $J_{1-dpcccg}$ and $J_{2-dpcccg}$

Obtain d_2 by Eq. (2.8) for $\vec{v}_g(5)$ and $\vec{v}_g(6)$ using respective $J_{1-dpcccg}$ and $J_{2-dpcccg}$

$J_{dpcccgK}[Sector] = J_{dpcccg}[5]d_1 + J_{dpcccg}[6]d_2$

end

case:6

Obtain d_1 by Eq. (2.8) for $\vec{v}_g(6)$ and $\vec{v}_g(1)$ using respective $J_{1-dpcccg}$ and $J_{2-dpcccg}$

Obtain d_2 by Eq. (2.8) for $\vec{v}_g(6)$ and $\vec{v}_g(1)$ using respective $J_{1-dpcccg}$ and $J_{2-dpcccg}$

$J_{dpcccgK}[Sector] = J_{dpcccg}[6]d_1 + J_{dpcccg}[1]d_2$

end

return $J_{dpcccgK}$

where H_1 is an $N_m \times N_c$ sparse matrix [98] and H_2 is an $N_m \times N_m$ dual diagonal square matrix; N_b, N_c, and N_m represent the number of control bits, coded bits, and parity bits, respectively.

The corresponding $N_b \times N_c$ systematic generator matrix G_{com} is given by

$$G_{com} = \begin{bmatrix} I & \psi \end{bmatrix}, \tag{6.11}$$

Algorithm 6.2 DPC FCS-MPC algorithm

while DPC FCS-MPC active **do**
  The system variables such as current, active and reactive power are measured
  $J_{op} = \infty$
  for
  h = 0 : 7 **do**
    $\vec{v}_{invg,\alpha\beta} = \vec{v}(h)$ % Possible states of the converter
    % System prediction of the grid current and power
    $\vec{i}_{g,\alpha\beta}(k+1) = \left(1 - \frac{TR_g}{L_g}\right)\vec{i}_{g,\alpha\beta}(k) + \frac{T}{L_g}(\vec{v}_{invg,\alpha\beta} - \vec{v}_{g,\alpha\beta}(k))$
    $i_{g,\alpha\ ref} = \frac{2(v_{g,\alpha}(k)P_{g\ ref} + v_{g,\beta}(k)Q_{g\ ref})}{3(v^2_{g,\alpha}(k) + v^2_{g,\beta}(k))}$
    $i_{g,\beta\ ref} = \frac{2(v_{g,\beta}(k)P_{g\ ref} - v_{g,\alpha}(k)Q_{g\ ref})}{3(v^2_{g,\alpha}(k) + v^2_{g,\beta}(k))}$
    $J_{dpcccg}[h] = |i_{g,\alpha\ ref} - i_{g,\alpha}(k+1)| + |i_{g,\beta\ ref} - i_{g,\beta}(k+1)|$ % Cost function calculation
  end for
  % Select the sector
  J_c = SectorFunction();
  for
  j = 0 : 7 **do**
    if $J_c(j) < J_{op}$ **then**
      $J_{op} = J_c(j)$
      $h_{op} = j$ % Store the sector with the least cost for a pair of vectors
    end if
  end for
end while
Apply the adjacent vector for respectively selected sector according to d_1 and d_2.

where I denotes the identity matrix, $\psi = H_1^T \times H_2^{-T}$, and H_2^{-T} is the upper-triangular matrix.

The decoding process can be performed using an iterative message-passing method [99,100]. In the encoding process, the control information vector $q_b = [q_b(0)\ \ldots\ q_b(N_b - 1)]^T$ is multiplied by the sparse matrix H_1^T and the resulting vector is differentially encoded to reach the parity bits. The systematic code-word vector $q_c = [q_c(0)\ \ldots\ q_c(N_c - 1)]^T$ is composed of the parity bits and the control bits.

To transmit the encoded information, the code-word vector is first interleaved and mapped to quaternary phase shift keying (QPSK) to improve the BER (bit error rate) performance, in addition to offering an adequate data rate

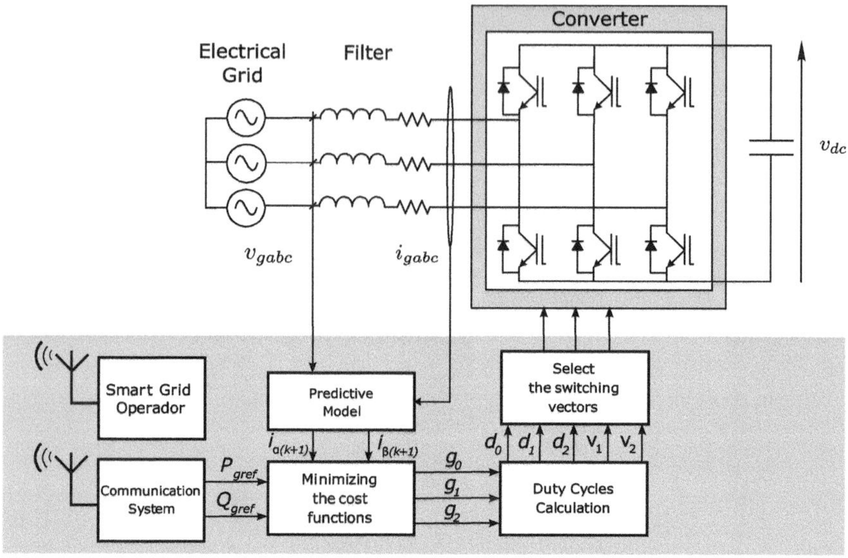

FIGURE 6.3 Block diagram for the modulated FCS-MPC DPC for CCG.

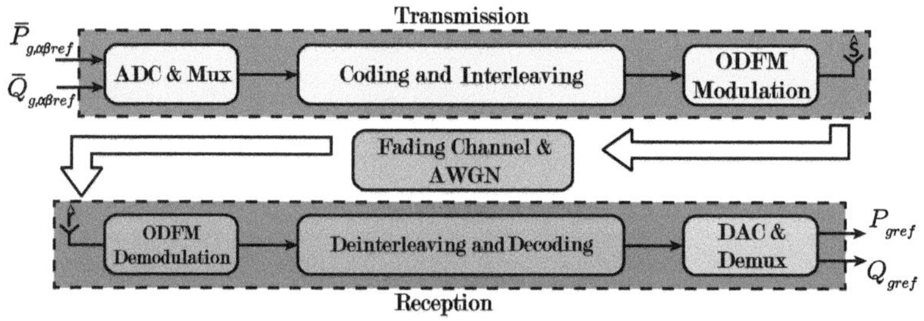

FIGURE 6.4 Block diagram for the wireless communication system.

[90]. Then, the resulting N_s QPSK symbols are divided into V subvectors of the N_f QPSK symbols $q_s^v = \left[q_s^v(0) \ \ldots \ q_s^v(N_f - 1)\right]^T$, where $0 < v < 1$, and applied to the orthogonal frequency division multiplexing (OFDM) process.

This transmission technique was chosen due to its high spectral efficiency. In this technique, the transmission and reception processes can be respectively performed by the inverse fast Fourier transform (IFFT) and fast Fourier transform (FFT) operations. In this way, the IFFT operation is applied to each of the V QPSK symbol subvectors, resulting in the respective OFDM symbol vectors $q_f^v = \left[q_f^v(0) \ \ldots \ q_f^v(N_f - 1)\right]^T$.

To obtain the final transmission symbol vectors $\vec{s} = \left[\vec{s}(-N_{cp}) \ \ldots \ \vec{s}(N_f - 1)\right]^T$, a cyclic prefix (CP) is added at the beginning of each OFDM symbol vector, in order to allow for the decoupling of transmissions received at the receiver in the frequency domain. The transmission is completed by assembling the symbol vectors, upconverting, filtering, and sending the information with an antenna.

At reception, as the CP duration is considered to be longer than the delay spread of the fading channel, after the removal of the CP symbols and the FFT operation, the equivalent low-pass discrete-time received vector, corresponding to each transmission symbol vector, can be represented by

$$\vec{r} = \vec{\jmath} \cdot \vec{s} + \vec{w}, \tag{6.12}$$

where $\vec{r} = \left[\vec{r}(0) \ldots \vec{r}(N_f - 1)\right]^T$ is the received vector, $\vec{\jmath} = \left[\vec{\jmath}(0) \ldots \vec{\jmath}(N_f - 1)\right]^T$ is the frequency response vector (frequency-selective fading channel), and $\vec{w} = \left[\vec{w}(0) \ldots \vec{w}(N_f - 1)\right]^T$ is additive white Gaussian noise (AWGN) vector in the frequency domain.

For each received vector, a one-tap least squares frequency domain equalization (LS-FDE) is applied and, then, the symbol demapping, code deinterleaving, and bit decoding operations are performed to recover the transmitted bits by the SG operator.

6.4 Analysis of the experimental results

Fig. 6.4 presents the wireless coded system which was simulated in MATLAB®/ Simulink, the equivalent baseband communication model was used as the basis for the analysis of an E_b/N_0 of 10 dB, representing a typical operation condition considered in all tests. A frequency-selective fading Rayleigh channel (with AWGN included) was used with a Doppler spread of 180 Hz (supposing an operation frequency of 2.45 GHz).

The wireless communication channel was varying in time (time-varying channel). On the other hand, the central control unit and renewable source were geographically fixed; to evaluate the proposed communication system, the worst scenario was chosen, the central control unit and the sources did not have line of sign (e.g., a wind farm or solar farm that can extend for kilometers) and the multipath components arrived at the receiver from all different directions with equal probability.

As performance reference, an ordinary CC scheme with high error correction capability was adopted, using a (171,133) generator polynomial and a constrain length of 7 [91]. Both CC and LDPC coding schemes had a 1/2 code rate and used random 64,800 length interleaving. The interactions were limited to 25 and the bit duration was 800 μs in the LDPC decoding scheme.

Each transmitted frame was composed of 36 OFDM symbols [101] (each one obtained by the IFT of a vector composed of 900 different coded QPSK symbols and 124 zero padding symbols) with the corresponding CP (each one obtained by the last 128 elements of the respective OFDM symbol vector), where N_f was 1024 (subchannels) and N_{cp} (cyclic prefix) was 128.

After the wireless communication processing, the received P_{gref} and Q_{gref} power references from the SG operator were inserted in the TMS320F28335 Texas Instruments digital signal processor (DSP), through a universal serial bus (USB) connection. Section 12.1.2 describes the setup, and the data is presented in Table B.1. Fig. 6.5[2] depicts the proposed experimental setup for the system.

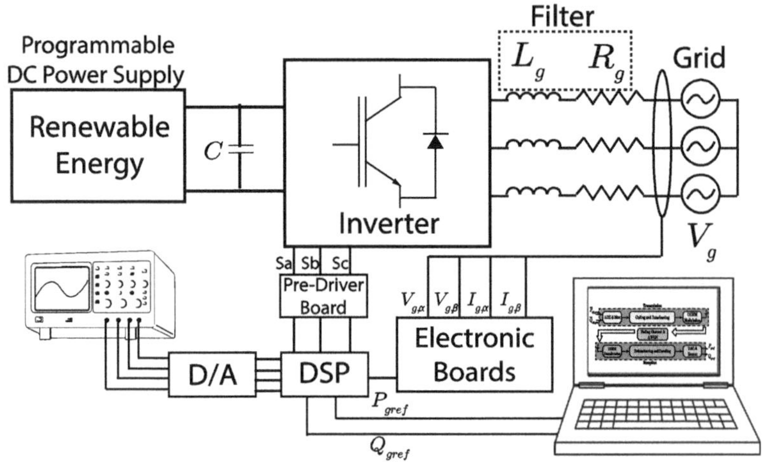

FIGURE 6.5 Experimental setup for tests.

6.4.1 OFDM-CC results

Fig. 6.6 depicts the test for OFDM communication, the test reference signal is for both power references (P_{gref} and Q_{gref}). The received active power changes from 500 to 750 W, and the received reactive power changes from −300 to 300 var. The good performance of the controller is checked again by evaluation that P_{gref} and Q_{gref} reference are reached by the controller even under distortion.

The response of the predictive controller employing OFDM modulation and CC scheme in the wireless link is presented in order to evaluate the OFDM process. Secondly, the test is a step up of the received P_{gref} from 500 to 750 W, it is executed each 100 ms, while Q_{gref} is maintained at −300 var. Fig. 6.7 depicts the test results where it can be observed that some spikes in the P_{gref} and Q_{gref} references occur. The result of errors in the wireless communication is shown by the spikes in the experimental response, due to the destructive effects of the channel, even if very efficient error correction scheme as CC is used.

The active power generated by the renewable source brings a significant variation around the reference, and in Fig. 6.7 it is noticeable that the grid current is distorted. Fig. 6.7 also presents the explicit error in the P_{gref} reference, as a spike around 70 ms, which drops the signal from 500 to 0 W, this error provides

[2] This figure is part of the work presented in [94].

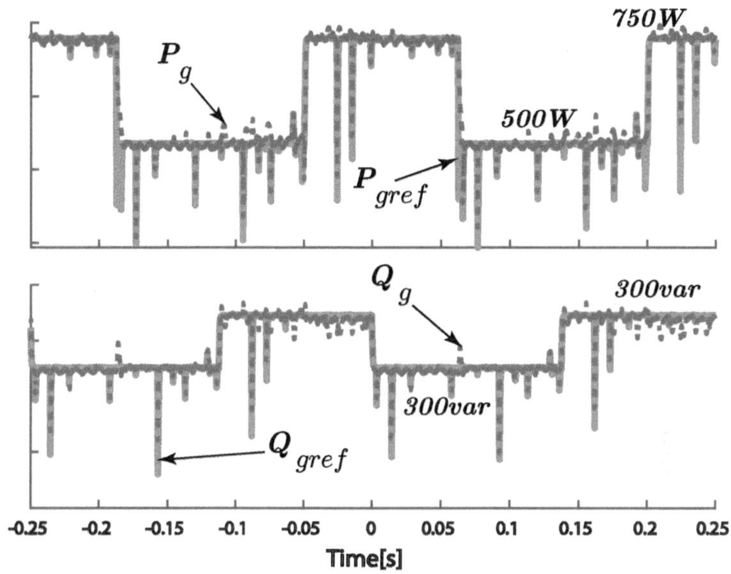

FIGURE 6.6 Response of received power step test, P_{gref} and Q_{gref}, using OFDM-CC.

an undesired P absorption by the grid. When observing the grid current signal, a notorious spike on the waveform can be seen, which generates unwanted harmonic components. It is necessary to highlight that a sudden variation in the current can cause unexpected conduction of IGBTs switches of the power converter, whose constant operation could damage the converter.

Fig. 6.8 introduces the current reference and current signal produced by the controller. It is noticeable that the reference has been tracked even if there are spikes due to the communication system.

Nonetheless, the performance of the whole system is not adequate for this SG application that involves OFDM wireless communication. This result provides evidence that it is mandatory to use a wireless system capable of minimizing the error, which is caused by the destructive characteristics of the channel. One of the solutions is the subject for the next section that shows a more robust wireless control system based on OFDM-LDPC, previously detailed in Section 6.3.

6.4.2 OFDM-LDPC results

The step test for both received power references is presented in Fig. 6.9, the active power changes from 500 to 750 W, and the reactive power changes from −300 to 300 var.

This test confirms the high performance of the controller according to the results presented below. Furthermore, the controller robustness can be observed

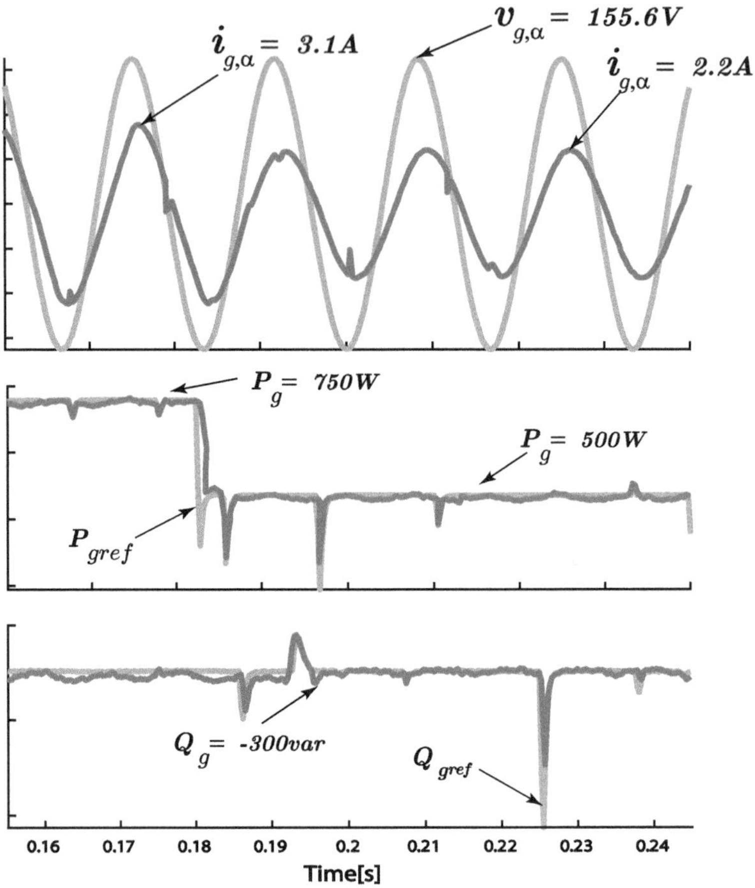

FIGURE 6.7 Grid voltage and current signals using OFDM-CC and response of received P_{gref} step test using OFDM-CC.

based on LDPC coding, since it is minimizing the occurrence of errors (spikes) in the recovered references of power.

For a real and fair comparison, all the tests were conducted with same steps and boundary conditions used from OFDM-CC tests. The referential signal for the second test is a step reference from 500 to 750 W for the P_{gref}, the Q_{gref} is maintained at −300 var, the result is presented in Fig. 6.10. The grid voltage and current are also shown in Fig. 6.10 with a time window of 100 ms (captured at the P maintained at 500 W).

A soft variation is noticed in the grid current, this soft grid current can avoid a lifetime reduction of IGBTs. Fig. 6.10 illustrates the better quality of the energy delivered to the grid when compared with the result shown in Fig. 6.7 demonstrating a satisfactory distortion reduction when an OFDM-LDPC scheme is used instead of OFDM-CC.

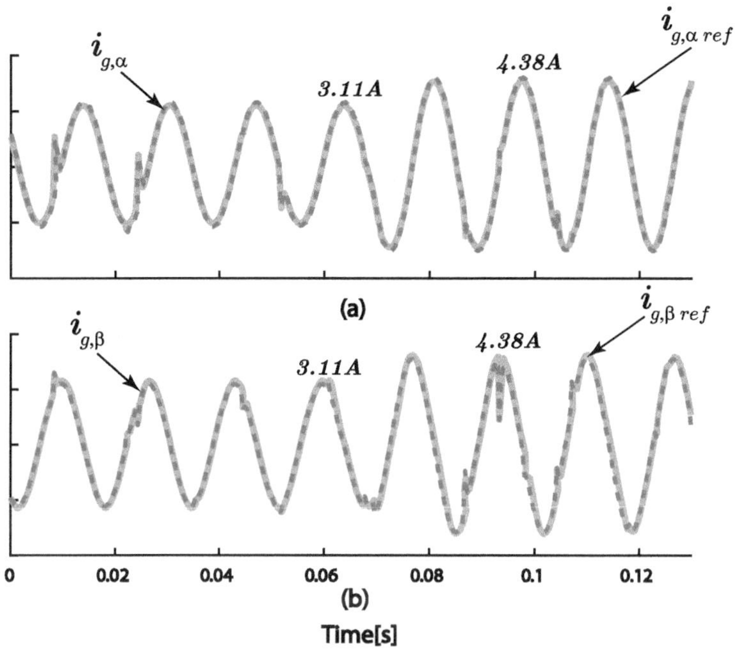

FIGURE 6.8 Grid current signals using OFDM-CC ($Q = -300$ var). The blue (gray in print version) line is the reference signal and the dashed red (dark gray in print version) line is the current signal.

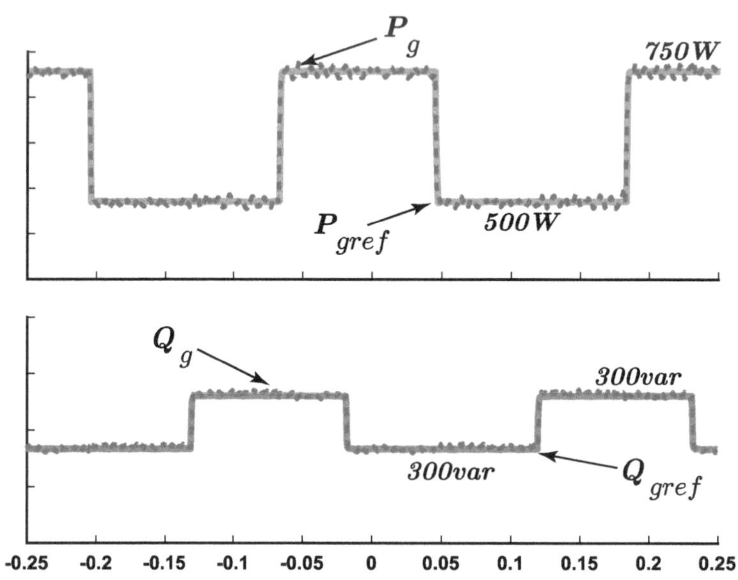

FIGURE 6.9 Response of received power step test, Q_{gref} and P_{gref}, using OFDM-LDPC.

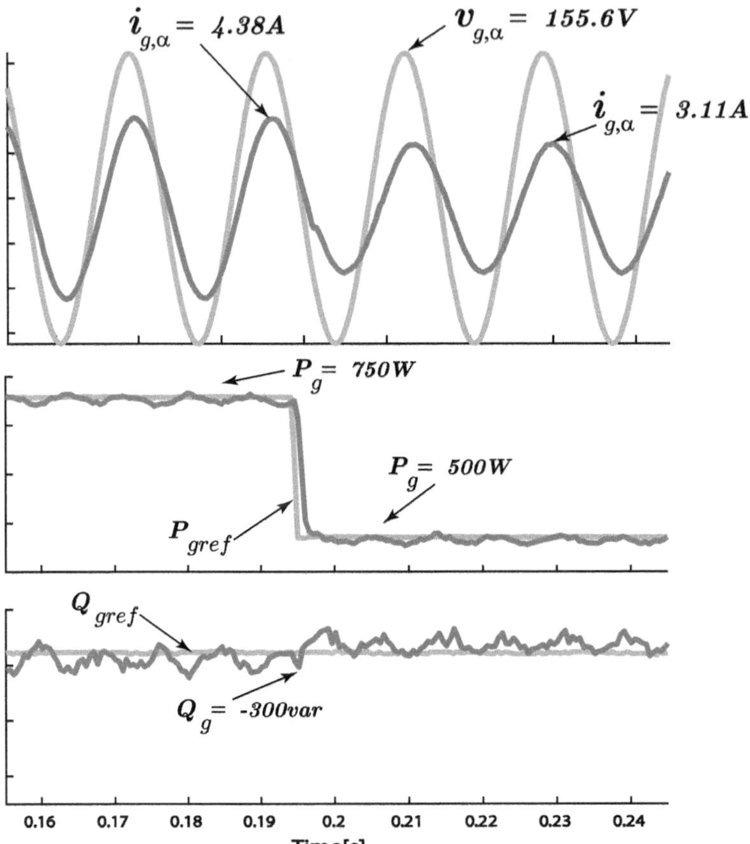

FIGURE 6.10 Grid voltage and current signals using OFDM-LDPC and response of received P_{gref} step test using OFDM-LDPC.

In fact, the absence of problematic spikes increases the performance of the system when using OFDM-LDPC when compared to the transmission using OFDM-CC, it is guaranteed for quality of energy produced by the smart grid system.

Fig. 6.11 presents the current reference and current signal produced by the controller; now both alpha and beta currents had an improvement when compared to Fig. 6.8.

6.4.3 Fast Fourier transform analysis

The controller applied in these tests is capable of operating at fixed switching frequency, in which Fig. 6.12 proves that the controller maintains the switching frequency of 10 kHz.

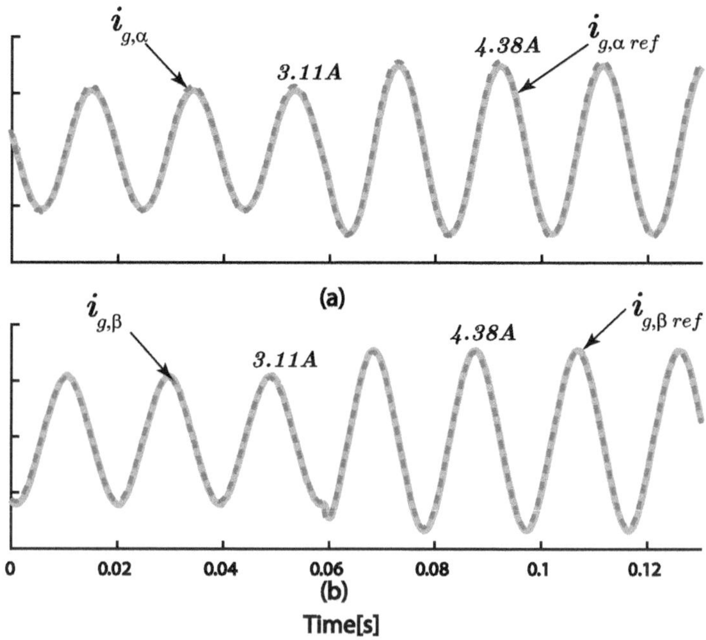

FIGURE 6.11 Grid current signals using OFDM-CC ($Q_{gref} = -300$ var).

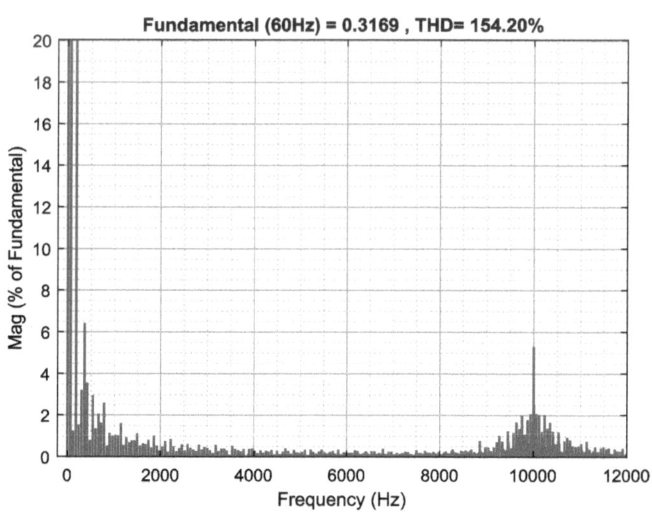

FIGURE 6.12 FFT analysis for switching.

Fig. 6.13 presents the THD for current in Fig. 6.11, where 1.69% represents the value according to IEEE STD 519-2014 [102] to certify the quality of current.

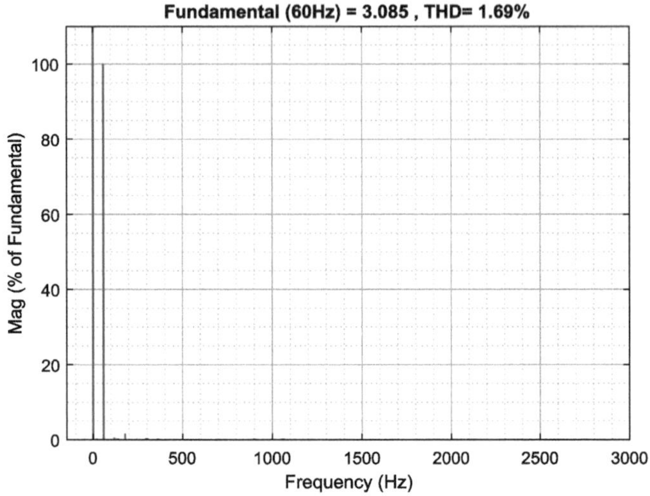

FIGURE 6.13 Current FFT analysis for OFDM-LDPC.

On the other hand, Fig. 6.14 presents the THD for current in Fig. 6.8, where 9.86% is a THD value above the limit required by IEEE STD 519-2014.

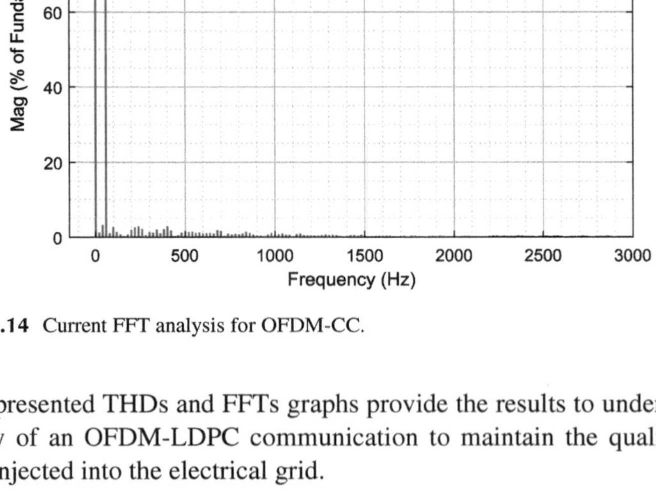

FIGURE 6.14 Current FFT analysis for OFDM-CC.

The presented THDs and FFTs graphs provide the results to understand the necessity of an OFDM-LDPC communication to maintain the quality of the current injected into the electrical grid.

6.5 Summary

Quality assurance for a power energy system based on the SG concept is directly related to the capability of the communication systems to maintain a reliable connection between operators and renewable sources, which is the future for energy generation.

As a contribution for the power energy area, it presents a proposal and experimental tests that connect a transmission/reception of power references to a renewable energy system through a wireless communication channel to remotely control a converter connected to the grid. The system proposed is based on the wireless communication using the OFDM-LDPC technique and the modulated FCS-MPC.

The OFDM with CC coding is used as a reference of performance, even if proportionately high signal-to-noise ratio is encountered compared to OFDM-LDPC. The OFDM-CC presents a lot of spikes, which implies a considerable variation in active and reactive power delivered to the grid. These variations are directly related to the transmission errors, which could affect the converter power components, at least deteriorate the energy quality injected into the grid.

This wireless power control system is employing an OFDM transmission scheme to avoid these impulsive spikes in both active power and reactive power references. This section presents the modern and future wireless communication systems, and an effective LDPC coding scheme. In order to improve the energy systems performance even under severe fading and noise conditions, which is increasing the quality of signal produced by decreasing or eliminating the error. Hence, the experimental analysis endorses the viability of the scheme due to the fact the power quality delivered to the power grid can be guaranteed.

Chapter 7

MPC-WM for doubly-fed induction generator and three-phase CCG

Alfeu J. Sguarezi Filho
Federal University of ABC, Santo André, SP, Brazil

7.1 DFIG rotor current control using MPC-WM

This section presents the rotor current control for DFIG by using MPC-WM theory introduced in Section 4.3 and in the work [103], in which just the simulation results were obtained. In this way, the DFIG can be represented by means of the rotor dynamic expressions in the dq reference frame due to the fact that the stator reactive and active power regulation can be done as per rotor current vector control presented in Section 3.1.1.

7.1.1 Space state equations

The rotor space state representation by using the rotor dynamics in the dq reference frame as presented in Eq. (3.28) can be represented as

$$\begin{aligned} px_{ir} &= A_{a,ir}x_{ir} + B_{b,ir}u_{u,ir} + G_{g,ir}d_{d,ir}, \\ y_{y,ir} &= C_{c,ir}x_{ir}. \end{aligned} \tag{7.1}$$

Hence, by using the rotor equations and considering the rotor current vector elements as the state, and considering the rotor voltage vector elements as the input, Eq. (7.1) can be stated as

$$\underbrace{\begin{bmatrix} \dfrac{di_{r,d}}{dt} \\ \dfrac{di_{r,q}}{dt} \end{bmatrix}}_{px_{ir}} = \underbrace{\begin{bmatrix} \dfrac{-R_r}{\sigma L_r} & \omega_{sl} \\ -\omega_{sl} & \dfrac{-R_r}{\sigma L_r} \end{bmatrix}}_{A_{a,ir}} \underbrace{\begin{bmatrix} i_{r,d} \\ i_{r,q} \end{bmatrix}}_{x_{ir}} + \underbrace{\begin{bmatrix} \dfrac{1}{\sigma L_r} & 0 \\ 0 & \dfrac{1}{\sigma L_r} \end{bmatrix}}_{B_{b,ir}} \underbrace{\begin{bmatrix} v_{r,d} \\ v_{r,q} \end{bmatrix}}_{u_{u,ir}}$$

https://doi.org/10.1016/B978-0-32-390964-8.00016-6

$$+\underbrace{\begin{bmatrix} 0 & \dfrac{\omega_{sl}L_m}{\sigma L_s L_r} \\ \dfrac{-\omega_{sl}L_m}{\sigma L_s L_r} & 0 \end{bmatrix}}_{G_{g,ir}} \underbrace{\begin{bmatrix} \lambda_s \\ 0 \end{bmatrix}}_{d_{d,ir}}, \tag{7.2}$$

where $p = \frac{d}{dt}$ and $C_{c,ir}$ is the identity matrix.

The rotor current vector control for DFIG by using MPC-WM employs the discretized state space equation considering the same assumption employed in Eq. (4.11). In this way, Eq. (7.2) can be discretized in a sampling time T by using the *zero-order-hold* (ZOH) technique without delay. Hence, Eqs. (7.1) and (7.2) can be expressed as

$$\begin{aligned} x_{ir}(k+1) &= A_{ad,ir}x_{ir}(k) + B_{bd,ir}u_{u,ir}(k) + G_{gd,ir}d_{d,ir}(k), \\ y_{ir}(k) &= C_{cd,ir}x_{ir}(k), \end{aligned} \tag{7.3}$$

where $A_{ad,ir} \cong I + A_{a,ir}T$, $B_{bd,ir} \cong B_{b,ir}T$, $G_{gd,ir} \cong G_{d,ir}T$, and $C_{cd,ir}$ is the identity matrix I. Hence, $y_{ir}(k) = x_{ir}(k)$. In this way, the first line of Eq. (7.3) can be stated as

$$\begin{aligned} \underbrace{\begin{bmatrix} i_{r,d}(k+1) \\ i_{r,q}(k+1) \end{bmatrix}}_{x_{ir}(k+1)} &= \underbrace{\begin{bmatrix} 1-\dfrac{R_r T}{\sigma L_r} & \omega_{sl}T \\ -\omega_{sl}T & 1-\dfrac{R_r T}{\sigma L_r} \end{bmatrix}}_{A_{ad,ir}} \underbrace{\begin{bmatrix} i_{r,d}(k) \\ i_{r,q}(k) \end{bmatrix}}_{x_{ir}(k)} \\ &+ \underbrace{\begin{bmatrix} \dfrac{T}{\sigma L_r} & 0 \\ 0 & \dfrac{T}{\sigma L_r} \end{bmatrix}}_{B_{bd,ir}} \underbrace{\begin{bmatrix} v_{r,d}(k) \\ v_{r,q}(k) \end{bmatrix}}_{u_{u,ir}(k)} \\ &+ \underbrace{\begin{bmatrix} 0 & \dfrac{\omega_{sl}L_M T}{\sigma L_s L_r} \\ \dfrac{-\omega_{sl}L_m T}{\sigma L_s L_r} & 0 \end{bmatrix}}_{G_{gd,ir}} \underbrace{\begin{bmatrix} \lambda_s(k) \\ 0 \end{bmatrix}}_{d_{d,ir}(k)}. \end{aligned} \tag{7.4}$$

In this chapter, the assumption is that ω_s and ω_r are constant at each sampling time. This can be achieved due to the electrical time constants being much minor than the mechanical time constants [104]. As the DFIG has normal operation conditions, $|\lambda_s|(k) = \lambda_s(k) = \frac{v_{s,q}(k)}{\omega_s(k)}$ is constant. Hence, the assumption for low values of n_y, $d_{d,ir}(k) \cong |\lambda_s|(k+1) \cong \cdots \cong |\lambda_s|(k+n_y-1)$ results in Eq. (4.20).

7.1.2 Rotor current control using MPC-WM

In this section the DFIG current control in the dq frame using MPC-WM is designed. One important piece of information in the MPC is the output prediction. In this way, the representation of the future rotor current reference in the dq frame by using Eqs. (4.17) and (7.4) can be stated as

$$YY_{ir} = AA_{ir}x_{ir}(k) + BB_{ir}\,VV_{ir} + GG_{ir}\,DD_{ir}, \tag{7.5}$$

where $VV_{ir} \in \mathbb{R}^{2n_u \times 1}$ is the prediction of inputs as rotor voltage vector components, $DD_{ir} \in \mathbb{R}^{2n_y \times 1}$ is the prediction of disturbance as stator flux magnitude, $YY_{ir} \in \mathbb{R}^{2n_y \times 1}$ is the prediction of outputs as rotor current vector components in the dq frame and,

$$YY_{ir} = \begin{bmatrix} y_{ir}(k+1) & y_{ir}(k+2) & y_{ir}(k+3) & \dots & y_{ir}(k+n_y) \end{bmatrix}^T, \tag{7.6}$$

$$VV_{ir} = \begin{bmatrix} u_{u,ir}(k) & u_{u,ir}(k+1) & u_{u,ir}(k+2) & \dots & u_{u,ir}(k+n_u-1) \end{bmatrix}^T, \tag{7.7}$$

$$\begin{aligned} DD_{ir} &= \begin{bmatrix} d_{d,ir}(k) & d_{d,ir}(k+1) & d_{d,ir}(k+2) & \dots & d_{d,ir}(k+n_y-1) \end{bmatrix}^T \\ &\approx d_{d,ir}(k)\begin{bmatrix} 1 & 1 & 1 & \dots & 1 \end{bmatrix}^T. \end{aligned} \tag{7.8}$$

Also, $AA_{ir} \in \mathbb{R}^{2n_y \times 2}$, $BB_{ir} \in \mathbb{R}^{2n_y \times 2n_u}$, and $GG_{ir} \in \mathbb{R}^{2n_y \times 2n_y}$ can be represented by Eqs. (7.9), (7.10), and (7.11), respectively:

$$AA_{ir} = \begin{bmatrix} A_{ad,ir} & A_{ad,ir}^2 & \dots & A_{ad,ir}^{n_y} \end{bmatrix}^T, \tag{7.9}$$

$$BB_{ir} = \begin{bmatrix} B_{bd,ir} & 0 & 0 & \dots \\ A_{ad,ir}B_{bd,ir} & B_{bd,ir} & 0 & \dots \\ A_{ad,ir}^2 B_{bd,ir} & A_{ad,ir}B_{bd,ir} & B_{bd,ir} & \dots \\ \vdots & \vdots & \vdots & \ddots \\ A_{ad,ir}^{n_y-1}B_{bd,ir} & A_{ad,ir}^{n_y-2}B_{bd,ir} & A_{ad,ir}^{n_y-3}B_{bd,ir} & \dots \end{bmatrix}, \tag{7.10}$$

$$GG_{ir} = \begin{bmatrix} G_{gd,ir} & 0 & 0 & \dots & 0 \\ A_{ad,ir}G_{gd,ir} & G_{gd,ir} & 0 & \dots & 0 \\ A_{ad,ir}^2 G_{gd,ir} & A_d G_{gd,ir} & G_{gd,ir} & \dots & 0 \\ \vdots & \vdots & \vdots & \ddots & \vdots \\ A_{ad,ir}^{n_y-1}G_{gd,ir} & A_{ad,ir}^{n_y-2}G_{gd,ir} & A_{ad,ir}^{n_y-3}G_{gd,ir} & \dots & G_{gd,ir} \end{bmatrix}. \tag{7.11}$$

The cost function without constraints can be represented by means of the future actions of the rotor current vector introduced in Eq. (7.5) as follows:

$$J_{ir} = (YY_{ir} - YY_{ref,ir})^T W_{yy,ir}(YY_{ir} - YY_{ref,ir}) + VV_{ir}^T W_{uu,ir} VV_{ir}, \quad (7.12)$$

where $W_{yy,ir} \in \mathbb{R}^{2n_y \times 2n_y}$ and $W_{uu,ir} \in \mathbb{R}^{2n_u \times 2n_u}$ represent the weighting matrix (Eqs. (4.28) and (4.29)); $YY_{ref,ir}$ represents the future references and it is constant at each sampling time. If necessary, the rotor current vector reference can be found applying the stator reactive and active power references by using Eqs. (3.33) and (3.34).

The minimized cost function by using Eq. (7.12) can be represented as

$$\begin{aligned} VV_{ir} = {} & (BB_{ir}^T W_{yy,ir} BB_{ir} + W_{uu,ir})^{-1} \\ & \cdot BB_{ir}^T W_{yy,ir}(YY_{ref,ir} - AA_{ir}x_x - GG_{ir}\, DD_{ir}). \end{aligned} \quad (7.13)$$

In this book, the rotor current of the DFIG is controlled by means of the voltage vector that is computed at each sampling time by the MPC-WM. Eq. (7.13) represents the elements of the rotor voltage vector ($\vec{v}_{r,dq}$) in the dq frame. Hence, the rotor voltage vector ($\vec{v}_{r,\alpha\beta r}$) in the $\alpha\beta r$ frame can be computed by using $\vec{v}_{r,\alpha\beta r} = \vec{v}_{r,dq} e^{j(\theta_s - \theta_r)}$, as presented in Eqs. (2.10)–(2.12). Vector $\vec{v}_{r,\alpha\beta r}$ can be the input of the PWM technique. So, the rotor current vector will reach the reference vector.

The DFIG rotor current control by using MPC-WM diagram is depicted in Fig. 7.1 and its basic code is presented in Algorithm 7.1.

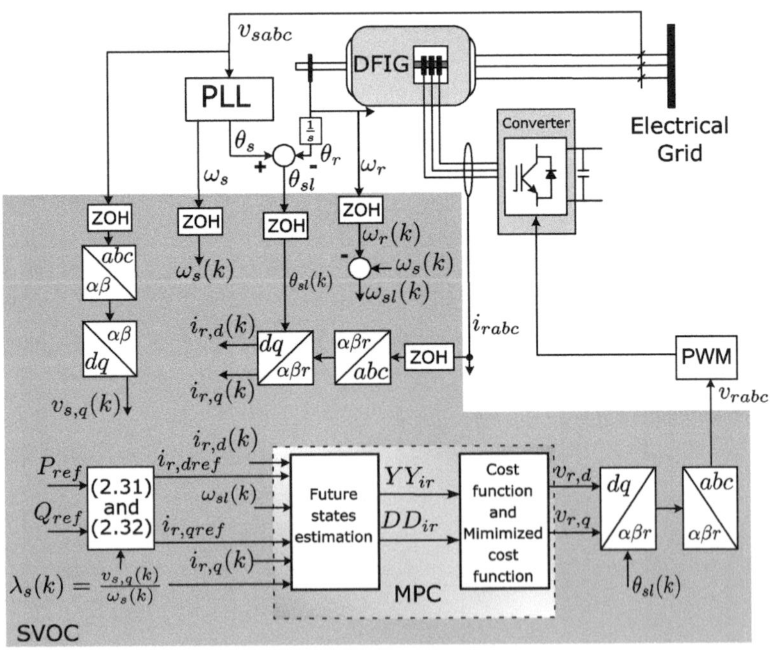

FIGURE 7.1 DFIG rotor current control by using MPC-WM diagram.

Algorithm 7.1 MPC-WM for DFIG rotor current control algorithm

while MPC-WM rotor current control active **do**
  $\vec{i}_{r,dq}(k), \omega_{sl}(k), \vec{i}_{r,dqref}(k) \leftarrow$ % System measurements
  if $\omega_{sl}(k) \neq \omega_{sl}(k-1)$ **then**
    $\omega_{sl}(k-1) = \omega_{sl}(k)$
    Calculate $A_{ad,ir}$, $B_{ad,ir}$, $G_{ad,ir}$, function of $(\omega_{sl}(k))$
    Calculate AA_{ir}, BB_{ir}, function of $(A_{ad,ir}, B_{ad,ir}, G_{ad,ir}, n_y, n_u)$
    Calculate GG_{ir}, DD_{ir}, function of $(A_{ad,ir}, B_{ad,ir}, G_{ad,ir}, n_y, n_u)$
  end if
  $YY_{ref,ir} \leftarrow$ Assign values of $\vec{i}_{r,dqref}$
  $W_{yy,ir}, W_{uu,ir} \leftarrow$ Assign values
  Calculate VV_{ir} by using Eq. (7.13)
  $\vec{v}_{r,\alpha\beta r} = \vec{v}_{r,dq} e^{j(\theta_s - \theta_r)}$
  Apply $\vec{v}_{r,\alpha\beta r}$ to the PWM technique.
end while

7.1.2.1 Simulation results

This section introduces the results obtained using computer simulations and the data presented in Table A.1. In this simulated test, the rotor speed is maintained at 1700 rpm. The horizons are $n_u = 2$, $n_y = 3$, $W_{yy,ir} = 2 \cdot 10^{10} I$, and $W_{uu,ir} = 3 \cdot 10^5 I$.

The behavior of the rotor current vector components in the dq frame during several step tests is depicted in Figs. 7.2(a)–(d). It can be noticed that the MPC-WM calculates the rotor voltage vector necessary for the rotor current vector components to reach their references and the rotor current performance during the test in: q – direct axis (Fig. 7.2(a)), d – quadrature axis (Fig. 7.2(b)), in the $\alpha\beta$ frame (Fig. 7.2(c)), and in the $\alpha\beta r$ frame (Fig. 7.2(d)).

7.1.2.2 Experimental results

This section introduces the experimental results obtained in the setup using the data presented in Table A.1, with the rotor speed maintained by the DC motor at 1700 rpm. The horizons are $n_u = 2$, $n_y = 3$, $W_{yy,ir} = 2 \cdot 10^{10} I$, and $W_{uu,ir} = 3 \cdot 10^5 I$.

The behavior of the rotor current vector components in the dq frame during several step tests is depicted in Fig. 7.3. It can be noticed that the MPC-WM calculates the rotor voltage vector necessary for the rotor current vector components to reach their references. Fig. 7.4 introduces the details of the mentioned test, in which the settling time is about 2 ms and $i_{r,d}$ remains constant. The oscillations occur due the switching of the inverter.

The performance of the stator current and voltage during the several step tests for the rotor current vector are depicted in Figs. 7.5 and 7.6, respectively. In both cases, (a) is related to the stator voltage and current and (b) is related to the rotor current vector components. It can be seen that the stator current increases

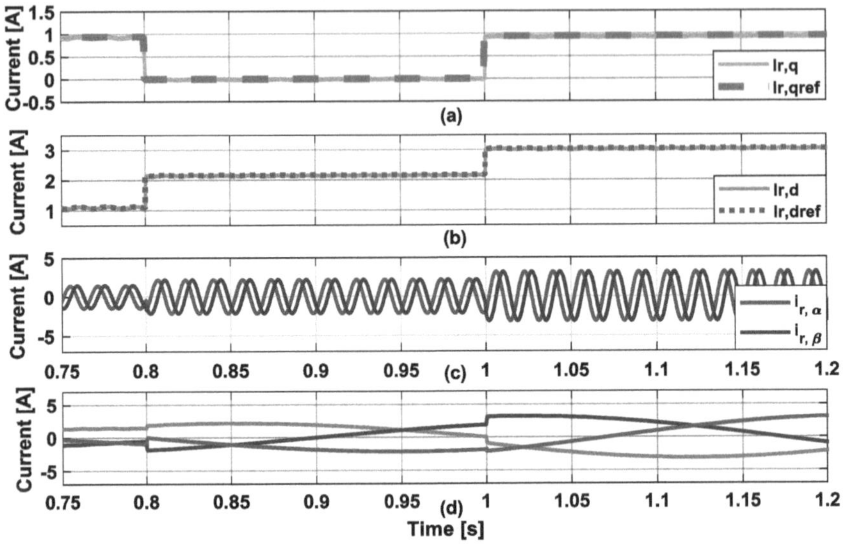

FIGURE 7.2 MPC-WM. Step tests for rotor current: (a) q – quadrature axis, (b) d – direct axis, (c) in the $\alpha\beta$ frame, and (d) in the three-phase rotor frame.

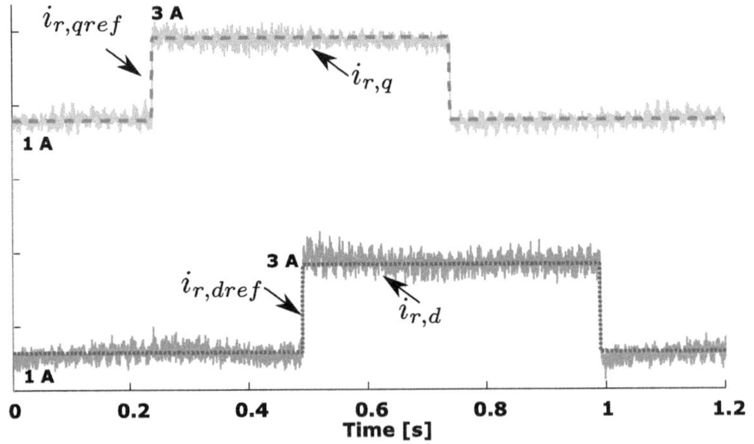

FIGURE 7.3 MPC-WM. Step tests for rotor current in the dq frame.

its amplitude when the $i_{r,q}$ changes its value, or the stator current changes its angle when $i_{r,d}$ changes its value due to the fact that the rotor current vector in the dq frame is related to the stator reactive and active power in accordance with the model representation.

Fig. 7.7 depicts a test when the speed of the DFIG changes from 1700 to 2000 rpm and then falls again to 1700 rpm smoothly, and the references of the rotor current vector are constant at $i_{r,q} = i_{r,d} = 1$ A. The actions of the

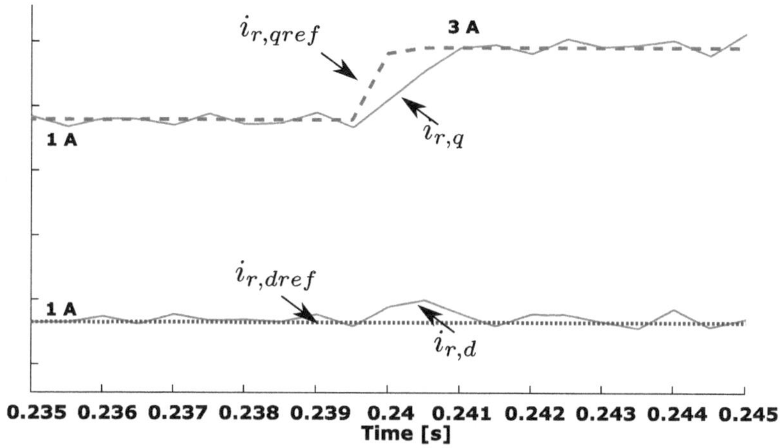

FIGURE 7.4 MPC-WM. Detail of step test for rotor current in the dq frame.

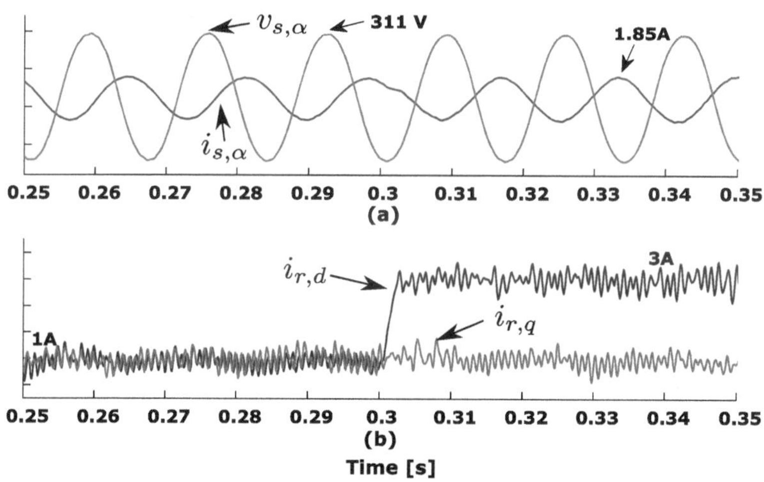

FIGURE 7.5 MPC-WM. Stator voltage and current during step of rotor current $i_{r,d}$: (a) stator voltage and current, (b) rotor current vector components.

speed of the rotor, the rotor current in rotor frame, and the rotor current vector components in the dq frame are depicted in Figs. 7.7(a)–(c), respectively. It can be noticed in Fig. 7.7 that the rotor current control by using MPC-WM reaches the references.

7.2 DFIG DPC using MPC-WM

This section presents the DPC for DFIG using MPC-WM theory presented in Section 4.3 and in work [105], in which just the simulation results are presented.

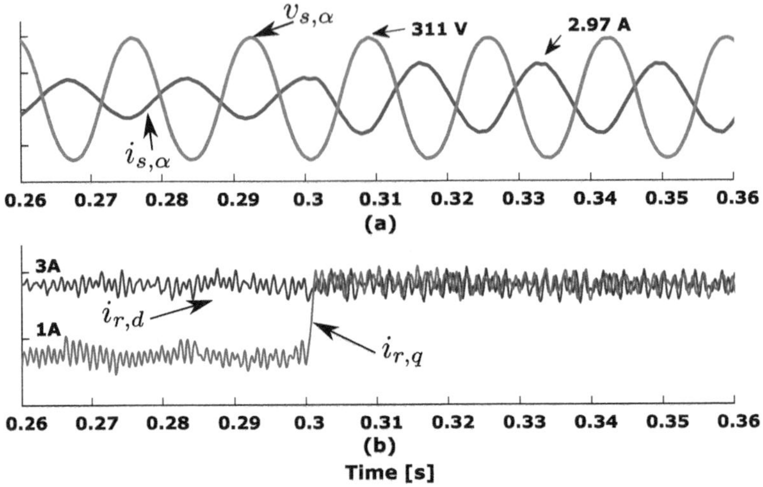

FIGURE 7.6 MPC-WM: Stator voltage and current during step of rotor current $i_{r,q}$. (a) Stator voltage and current. (b) Rotor current vector components.

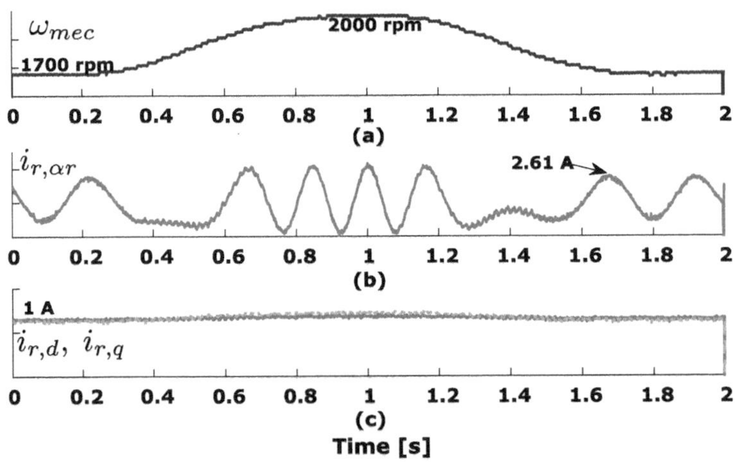

FIGURE 7.7 MPC-WM: Several speed operation test. (a) Speed of the rotor. (b) Rotor current in rotor frame. (c) Rotor current vector components in dq frame.

In this way, in the DPC MPC-WM, the controller manipulates directly the stator reactive and active power by means of the rotor voltage vector elements. The rotor space state representation using the rotor dynamics in the dq frame as per Eqs. (3.39) and (3.40) can be represented as

$$
\begin{aligned}
px_{pq} &= A_{a,pq}x_{pq} + B_{b,pq}u_{u,pq} + G_{g,pq}d_{d,pq}, \\
y_{y,pq} &= C_{c,pq}x_{pq},
\end{aligned}
\tag{7.14}
$$

where

$$\underbrace{\begin{bmatrix} \dfrac{dQ_s}{dt} \\ \dfrac{dP_s}{dt} \end{bmatrix}}_{px_{pq}} = \underbrace{\begin{bmatrix} 0 & \omega_{sl} \\ -\omega_{sl} & 0 \end{bmatrix}}_{A_{a,pq}} \underbrace{\begin{bmatrix} Q_s \\ P_s \end{bmatrix}}_{x_{pq}} + \underbrace{\begin{bmatrix} \frac{1}{Am} & 0 \\ 0 & \frac{1}{Am} \end{bmatrix}}_{B_{b,pq}} \underbrace{\begin{bmatrix} v_{r,d} \\ v_{r,q} \end{bmatrix}}_{u_{u,pq}} + \underbrace{\begin{bmatrix} 1 & 0 \\ 0 & 1 \end{bmatrix}}_{G_{g,pq}} \underbrace{\begin{bmatrix} 0 \\ -\frac{\omega_{sl} L_r}{L_m\, Am} \lambda_s \end{bmatrix}}_{d_{d,pq}} \tag{7.15}$$

and $p = \frac{d}{dt}$, $C_{c,pq} = I$, with I being the identity matrix, and $Am = -\frac{2\sigma L_s L_r}{3 v_s L_m}$.

The DPC for DFIG by using MPC-WM employs the discrete representation of the state space equation in which the input is represented by means of the rotor voltage vector and the state is represented as per the stator reactive and active power. In this way, Eq. (7.15) can be discretized in a sampling period T by means of the *zero-order-hold* (ZOH) technique without delay. Hence, Eqs. (7.14) and (7.15) can be rewritten as follows:

$$\begin{aligned} x_{pq}(k+1) &= A_{ad,pq} x_{pq}(k) + B_{bd,pq} u_{u,pq}(k) + G_{gd,pq} d_{d,pq}(k), \\ y_{pq}(k) &= C_{cd,pq} x_{pq}(k), \end{aligned} \tag{7.16}$$

where $A_{ad,pq} \cong I + A_{a,pq} T$, $B_{bd,pq} \cong B_{b,ir} T$, $G_{gd,pq} \cong G_{d,pq} T$, and $C_{cd,pq}$ is the identity matrix I. In this way, the first line of Eq. (7.16) can be stated as:

$$\underbrace{\begin{bmatrix} Q_s(k+1) \\ P_s(k+1) \end{bmatrix}}_{x_{pq}(k+1)} = \underbrace{\begin{bmatrix} 1 & \omega_{sl} T \\ -\omega_{sl} T & 1 \end{bmatrix}}_{A_{ad,pq}} \underbrace{\begin{bmatrix} Q_s(k) \\ P_s(k) \end{bmatrix}}_{x_{pq}(k)} + \underbrace{\begin{bmatrix} \frac{T}{Am} & 0 \\ 0 & \frac{T}{Am} \end{bmatrix}}_{B_{bd,pq}} \underbrace{\begin{bmatrix} v_{r,d}(k) \\ v_{r,q}(k) \end{bmatrix}}_{u_{u,pq}(k)} + \underbrace{\begin{bmatrix} 1 & 0 \\ 0 & 1 \end{bmatrix}}_{G_{gd,pq}} \underbrace{\begin{bmatrix} 0 \\ \frac{-\omega_{sl} L_r T}{L_M\, Am} \lambda_s(k) \end{bmatrix}}_{d_{d,pq}(k)}. \tag{7.17}$$

As stated in the last section, the assumption is that ω_s and ω_r are constant at each sampling time. This can be done due to the electrical time constants being much minor than the mechanical time constants [104]. As the DFIG has normal operation conditions, $|\lambda_s|(k) = \lambda_s(k) = \frac{v_{s,q}(k)}{\omega_s(k)}$ is constant. Hence, the assumption for low values of n_y, $d_{d,pq}(k) \cong \frac{-\omega_{sl} L_2 T}{L_M\, Am} \lambda_s(k+1) \cong \cdots \cong \frac{-\omega_{sl} L_2 T}{L_M\, Am} \lambda_s(k + n_y - 1)$ results in Eq. (4.20).

7.2.1 DPC using MPC-WM

In this section the DPC for DFIG using MPC-WM is designed. The representation of the future stator reactive and active power by using Eqs. (4.17) and (7.17) can be expressed as follows:

$$YY_{pq} = AA_{pq} x_{pq}(k) + BB_{pq}\; VV_{pq} + GG_{pq}\; DD_{pq}, \tag{7.18}$$

where $VV_{pq} \in \mathbb{R}^{2n_u \times 1}$ represents the prediction of inputs as rotor voltage vector components, $DD_{pq} \in \mathbb{R}^{2n_y \times 1}$ expresses the prediction of disturbance as stator flux magnitude, $YY_{pq} \in \mathbb{R}^{2n_y \times 1}$ represents the prediction of outputs components as stator active and reactive power, and

$$YY_{pq} = \begin{bmatrix} y_{pq}(k+1) & y_{pq}(k+2) & y_{pq}(k+3) & \ldots & y_{pq}(k+n_y) \end{bmatrix}^T, \tag{7.19}$$

$$VV_{pq} = \begin{bmatrix} u_{u,pq}(k) & u_{u,pq}(k+1) & u_{u,pq}(k+2) & \ldots & u_{u,pq}(k+n_u-1) \end{bmatrix}^T, \tag{7.20}$$

$$\begin{aligned} DD_{pq} &= \begin{bmatrix} d_{d,pq}(k) & d_{d,pq}(k+1) & d_{d,pq}(k+2) & \ldots & d_{d,pq}(k+n_y-1) \end{bmatrix}^T \\ &\approx d_{d,pq}(k) \begin{bmatrix} 1 & 1 & 1 & \ldots & 1 \end{bmatrix}^T. \end{aligned} \tag{7.21}$$

Also, $AA_{pq} \in \mathbb{R}^{2n_y \times 2}$, $BB_{pq} \in \mathbb{R}^{2n_y \times 2n_u}$, and $GG_{pq} \in \mathbb{R}^{2n_y \times 2n_y}$ can be represented by using Eqs. (7.22), (7.23) and (7.24), respectively:

$$AA_{pq} = \begin{bmatrix} A_{ad,pq} & A_{ad,pq}^2 & \cdots & A_{ad,pq}^{n_y} \end{bmatrix}^T, \tag{7.22}$$

$$BB_{pq} = \begin{bmatrix} B_{bd,pq} & 0 & 0 & \cdots \\ A_{ad,pq}B_{bd,pq} & B_{bd,pq} & 0 & \cdots \\ A_{ad,pq}^2 B_{bd,pq} & A_{ad,pq}B_{bd,pq} & B_{bd,pq} & \cdots \\ \vdots & \vdots & \vdots & \ddots \\ A_{ad,pq}^{n_y-1} B_{bd,pq} & A_{ad,pq}^{n_y-2} B_{bd,pq} & A_{ad,pq}^{n_y-3} B_{bd,pq} & \cdots \end{bmatrix}, \tag{7.23}$$

$$GG_{pq} = \begin{bmatrix} G_{gd,pq} & 0 & 0 & \cdots & 0 \\ A_{ad,pq}G_{gd,pq} & G_{gd,pq} & 0 & \cdots & 0 \\ A_{ad,pq}^2 G_{gd,pq} & A_d G_{gd,pq} & G_{gd,pq} & \cdots & 0 \\ \vdots & \vdots & \vdots & \ddots & \vdots \\ A_{ad,pq}^{n_y-1} G_{gd,pq} & A_{ad,pq}^{n_y-2} G_{gd,pq} & A_{ad,pq}^{n_y-3} G_{gd,pq} & \cdots & G_{gd,pq} \end{bmatrix}. \tag{7.24}$$

The cost function without constraints can be represented by means of the future behavior of the stator reactive and active power as introduced in Eq. (7.18), and one can express it as follows:

$$J_{pq} = (YY_{pq} - YY_{ref,pq})^T W_{yy,pq}(YY_{pq} - YY_{ref,pq}) + VV_{pq}^T W_{uu,pq} VV_{pq}, \tag{7.25}$$

where $W_{yy,pq} \in \mathbb{R}^{2n_y \times 2n_y}$ and $W_{uu,pq} \in \mathbb{R}^{2n_u \times 2n_u}$ represent the weighting matrices (Eqs. (4.28) and (4.29)); $YY_{ref,pq}$ represents the future references of stator reactive and active power, and they have the same value at each sampling time.

The minimized cost function by using Eq. (7.25) can be represented as

$$\begin{aligned} VV_{pq} = {} & (BB_{pq}^T W_{yy,pq} BB_{pq} + W_{uu,pq})^{-1} \\ & \cdot BB_{pq}^T W_{yy,pq}(YY_{ref,pq} - AA_{pq}x_x - GG_{pq}\, DD_{pq}). \end{aligned} \tag{7.26}$$

Eq. (7.26) represents the elements of the rotor voltage vector ($\vec{v}_{r,dq}$) in the dq frame. Hence, the rotor voltage vector ($\vec{v}_{r,\alpha\beta r}$) in the $\alpha\beta r$ frame can be calculated by using $\vec{v}_{r,\alpha\beta r} = \vec{v}_{r,dq} e^{j(\theta_s - \theta_r)}$ as presented in Eqs. (2.10)–(2.12). Vector $\vec{v}_{r,\alpha\beta r}$ can be the input of the space vector PWM. Thus, the stator apparent power will reach each element of the stator power references.

The DFIG DPC by using MPC-WM is depicted in Fig. 7.8, and its basic code is presented in Algorithm 7.2.

7.2.1.1 Experimental results

This section introduces the experimental results obtained in the setup using the data presented in Table A.1, and the rotor speed is maintained by the DC motor at 1650 rpm. The horizons are $n_u = 2$, $n_y = 3$, $W_{yy,pq} = 4 \cdot 10^{10} I$, and $W_{uu,pq} = 7 \cdot 10^5 I$.

The behavior of the stator reactive and active power over several step tests is depicted in Fig. 7.9, in which the active power changes from −350 to −1050 W and the reactive power changes from −1050 to −350 var. It can be noticed that the DPC for DFIG using MPC-WM calculates the rotor voltage vector necessary for the stator power to reach its references. The settling time is about 2 ms as can be seen in Fig. 7.10, and in this test the reactive power (Q) remains constant. The oscillations occur due the switching of the inverter.

The actions of the stator current and voltage during several step tests for the power references are depicted in Figs. 7.11 and 7.12, respectively. In both cases, (a) is related to the stator power and (b) is related to the stator voltage and current. It can be seen that the stator current increases its amplitude when the active power changes its value, or the stator current changes its angle when the reactive power changes its value due to the fact that the stator current vector in the dq frame is related to the stator active and reactive power in accordance with the model representation.

In the last test, DFIG changes its speed from 2000 to 1700 rpm, the stator reactive power is maintained at −1050 var, and the stator active power is maintained at −350 W. The actions of the speed of the rotor, the rotor current in the rotor frame, and the stator power elements are depicted in Figs. 7.13(a)–(c),

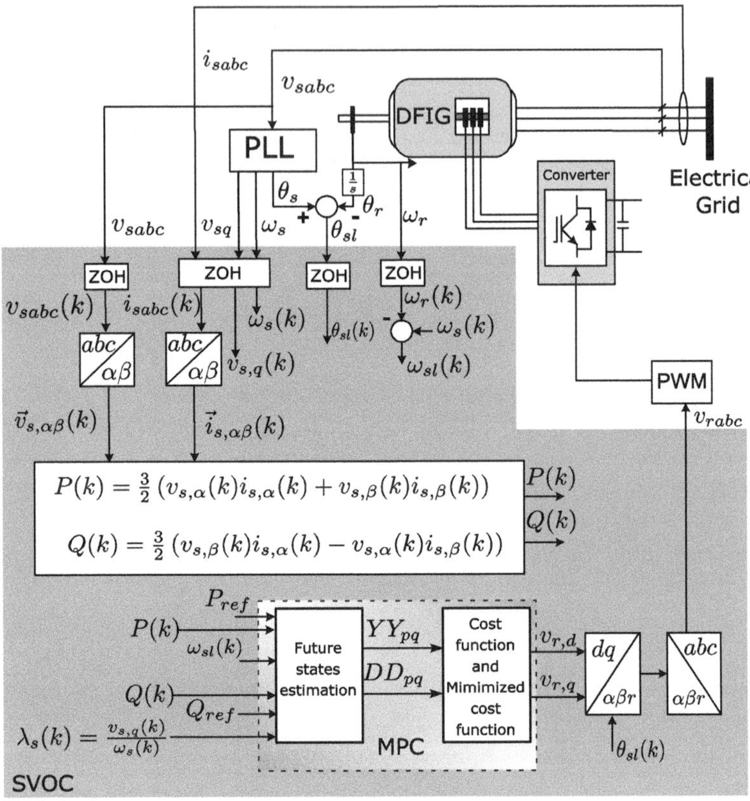

FIGURE 7.8 MPC-WM: DFIG DPC.

Algorithm 7.2 DPC MPC-WM for DFIG algorithm

while DPC MPC-WM active **do**

$\vec{v}_{s,dq}(k), \omega_{sl}(k), P(k), Q(k), P_{ref}(k), Q_{ref}(k) \leftarrow$ % System measurements

if $\omega_{sl}(k) \neq \omega_{sl}(k-1)$ **then**

$\omega_{sl}(k-1) = \omega_{sl}(k)$

Calculate $A_{ad,pq}$, $B_{ad,pq}$, $G_{ad,pq}$, function of $(\omega_{sl}(k))$

Calculate AA_{pq}, BB_{pq}, function of $(A_{ad,pq}, B_{ad,pq}, G_{ad,pq}, n_y, n_u)$

Calculate GG_{pq}, DD_{pq}, function of $(A_{ad,pq}, B_{ad,pq}, G_{ad,pq}, n_y, n_u)$

end if

$YY_{ref,pq} \leftarrow$ Assign values of $\vec{i}_{r,dqref}$

$W_{yy,pq}, W_{uu,pq} \leftarrow$ Assign values

Calculate VV_{pq} by using Eq. (7.26)

$\vec{v}_{r,\alpha\beta r} = \vec{v}_{r,dq} e^{j(\theta_s - \theta_r)}$

Apply $\vec{v}_{r,\alpha\beta r}$ to the PWM technique.

end while

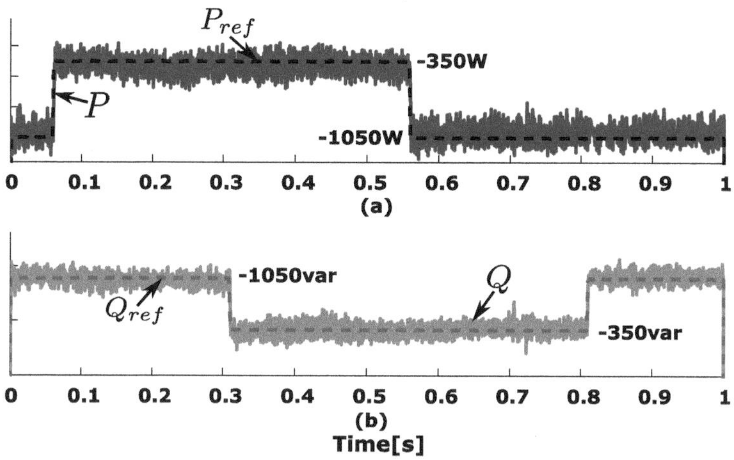

FIGURE 7.9 DPC MPC-WM. Step tests for active and reactive power: (a) active power, (b) reactive power.

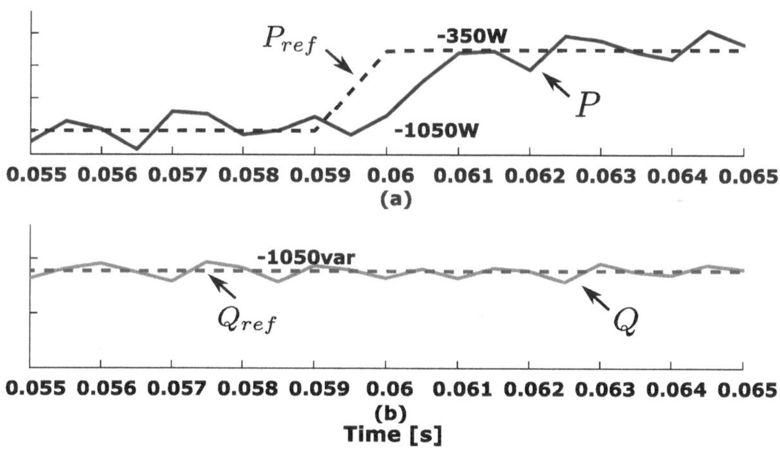

FIGURE 7.10 DPC MPC-WM. Detail of step test for active power: (a) active power, (b) reactive power.

respectively. It can be noticed in Fig. 7.13 that the stator power components control by using MPC-WM reaches the references.

7.3 Three-phase CCG current control using MPC-WM

This section presents the three-phase CCG current control using MPC-WM theory presented in Section 4.3 operating in the $\alpha\beta$ frame and the theory was introduced in [106]. In this way, the grid reactive and active power can be regulated as per the control of the elements of the grid current vector.

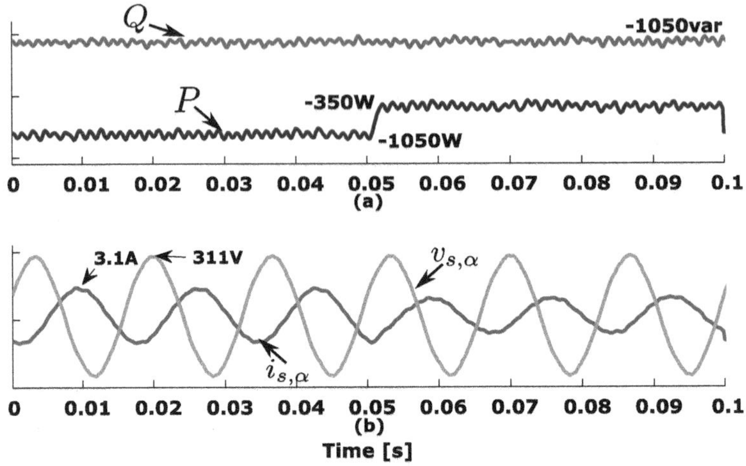

FIGURE 7.11 DPC MPC-WM. Stator voltage and current during step of active power: (a) reactive and active power, (b) stator voltage and current.

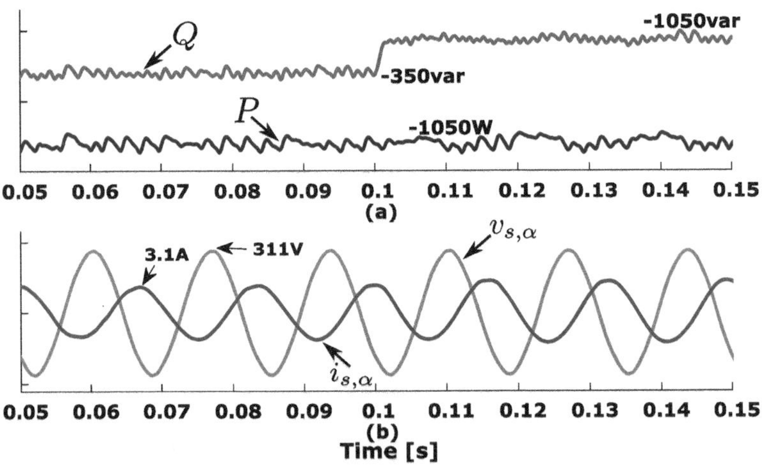

FIGURE 7.12 DPC MPC-WM. Stator voltage and current during step of reactive power: (a) reactive and active power, (b) stator voltage and current.

7.3.1 Space state equations

The CCG space state in the $\alpha\beta$ frame can be expressed by using Eqs. (2.64)–(2.66) as

$$\begin{aligned} px_g &= A_{a,g}x_g + B_{b,g}u_{u,g}, \\ y_{y,g} &= C_{c,g}x_g. \end{aligned} \tag{7.27}$$

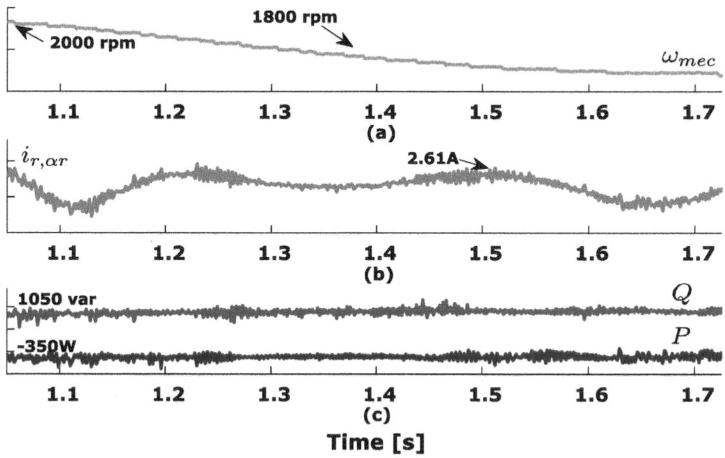

FIGURE 7.13 DPC MPC-WM. Several-speed operation test: (a) rotor speed, (b) rotor current, (c) reactive and active power.

Hence, the first line of Eq. (7.27) can be stated as follows:

$$\underbrace{\begin{bmatrix} \dfrac{di_{g,\alpha}}{dt} \\ \dfrac{di_{g,\beta}}{dt} \end{bmatrix}}_{px_g} = \underbrace{\begin{bmatrix} \dfrac{-R_g}{L_g} & 0 \\ 0 & \dfrac{-R_g}{L_g} \end{bmatrix}}_{A_{a,g}} \underbrace{\begin{bmatrix} i_{g,\alpha} \\ i_{g,\beta} \end{bmatrix}}_{x_g} + \underbrace{\begin{bmatrix} \dfrac{1}{L_g} & 0 \\ 0 & \dfrac{1}{L_g} \end{bmatrix}}_{B_{b,g}} \underbrace{\begin{bmatrix} v_{inv,\alpha} - v_{g,\alpha} \\ v_{inv,\beta} - v_{g,\beta} \end{bmatrix}}_{u_{u,g}}, \quad (7.28)$$

where $p = \frac{d}{dt}$ and $C_{c,g}$ is the identity matrix.

The grid current control for CCG by using MPC-WM employs the discretized state space equation, in which the grid current vector is the state and the converter voltage vector acts as the input. In this way, Eq. (7.28) can be discretized in a sampling time T by employing the *zero-order-hold* (ZOH) technique without delay. Hence, Eqs. (7.27) and (7.28) can be represented as follows:

$$\begin{aligned} x_g(k+1) &= A_{ad,g} x_g(k) + B_{bd,g} u_{u,g}(k), \\ y_g(k) &= C_{cd,g} x_g(k), \end{aligned} \quad (7.29)$$

where $A_{ad,g} \cong I + A_{a,g}T$, $B_{bd,g} \cong B_{b,g}T$, and $C_{cd,g}$ is the identity matrix I. Thus, the first line of Eq. (7.29) can be stated as follows:

$$\underbrace{\begin{bmatrix} i_{g,\alpha}(k+1) \\ i_{g,\beta}(k+1) \end{bmatrix}}_{x_g(k+1)} = \underbrace{\begin{bmatrix} 1 - \dfrac{R_g T}{L_g} & 0 \\ 0 & 1 - \dfrac{R_g T}{L_g} \end{bmatrix}}_{A_{ad,g}} \underbrace{\begin{bmatrix} i_{g,\alpha}(k) \\ i_{g,\beta}(k) \end{bmatrix}}_{x_g(k)}$$

$$+\underbrace{\begin{bmatrix}\dfrac{T}{L_g} & 0\\ 0 & \dfrac{T}{L_g}\end{bmatrix}}_{B_{bd,g}}\underbrace{\begin{bmatrix}v_{inv,\alpha}(k)-v_{g,\alpha}(k)\\ v_{inv,\beta}(k)-v_{g,\beta}(k)\end{bmatrix}}_{u_{u,g}(k)}. \qquad (7.30)$$

As the electrical grid is under normal operation conditions, the assumption for low values of n_y, $\vec{v}_{g,\alpha\beta}(k) \cong \vec{v}_{g,\alpha\beta}(k+1) \cong \cdots \cong \vec{v}_{g,\alpha\beta}(k+n_y-1)$ is due the sampling time T being much smaller than the period of the grid time.

7.3.2 Grid current control using MPC-WM

In this section the grid current regulation using MPC-WM is introduced. The representation of the future grid current by using Eqs. (4.17) and (7.30) can be stated as follows:

$$YY_g = AA_g x_g(k) + BB_g\, VV_g, \qquad (7.31)$$

where $VV_g \in \mathbb{R}^{2n_u\times 1}$ expresses the prediction of inputs as CCG minus grid voltage vectors components, $YY_g \in \mathbb{R}^{2n_y\times 1}$ represents the prediction of outputs as grid current vector components, and

$$YY_g = \begin{bmatrix} y_g(k+1) & y_g(k+2) & y_g(k+3) & \dots & y_g(k+n_y)\end{bmatrix}^T, \qquad (7.32)$$

$$VV_g = \begin{bmatrix} u_{u,g}(k) & u_{u,g}(k+1) & u_{u,g}(k+2) & \dots & u_{u,g}(k+n_u-1)\end{bmatrix}^T. \qquad (7.33)$$

Also, $AA_g \in \mathbb{R}^{2n_y\times 2}$, $BB_g \in \mathbb{R}^{2n_y\times 2n_u}$, and $GG_g \in \mathbb{R}^{2n_y\times 2n_y}$ can be represented by Eqs. (7.34) and (7.35), respectively:

$$AA_g = \begin{bmatrix} A_{ad,g} & A_{ad,g}^2 & \dots & A_{ad,g}^{n_y}\end{bmatrix}^T, \qquad (7.34)$$

$$BB_g = \begin{bmatrix} B_{bd,g} & 0 & 0 & \dots\\ A_{ad,g}B_{bd,g} & B_{bd,g} & 0 & \dots\\ A_{ad,g}^2 B_{bd,g} & A_{ad,g}B_{bd,g} & B_{bd,g} & \dots\\ \vdots & \vdots & \vdots & \ddots\\ A_{ad,g}^{n_y-1}B_{bd,g} & A_{ad,g}^{n_y-2}B_{bd,g} & A_{ad,g}^{n_u-3}B_{bd,g} & \dots \end{bmatrix}. \qquad (7.35)$$

The cost function without constraints can be represented by means of the future behavior of the grid current presented in Eq. (7.31) as

$$J_g = (YY_g - YY_{ref,g})^T W_{yy,g}(YY_g - YY_{ref,g}) + VV_g^T W_{uu,g} VV_g, \qquad (7.36)$$

where $W_{yy,g} \in \mathbb{R}^{2n_y \times 2n_y}$ and $W_{uu,g} \in \mathbb{R}^{2n_u \times 2n_u}$ represent the weighting matrices (Eqs. (4.28) and (4.29)); $YY_{ref,g}$ represents the future references of the grid current, and they are constant at each sampling time. The references of the grid current elements can be computed by means of the references of the grid reactive and active power as presented in [43,107]:

$$\begin{bmatrix} i_{g,\alpha ref} \\ i_{g,\beta ref} \end{bmatrix} = \frac{2}{3[v_{g,\alpha}^2(k) + v_{g,\beta}^2(k)]} \begin{bmatrix} v_{g,\alpha}(k) & v_{g,\beta}(k) \\ v_{g,\beta}(k) & -v_{g,\alpha}(k) \end{bmatrix} \begin{bmatrix} P_{gref} \\ Q_{gref} \end{bmatrix}. \quad (7.37)$$

The minimized cost function by using Eq. (7.36) can be written as

$$VV_g = (BB_g^T W_{yy,g} BB_g + W_{uu,g})^{-1} BB_g^T W_{yy,g} (YY_{ref,g} - AA_g x_g). \quad (7.38)$$

Eq. (7.38) represents the components of the converter minus voltage vector the grid voltage vector $(\vec{v}_{inv,\alpha\beta} - \vec{v}_{g,\alpha\beta}(k))$ in the $\alpha\beta$ frame. Hence, vector $(\vec{v}_{inv,\alpha\beta})$ can be the input of the space vector PWM. So, the elements of the grid current vector will converge to the references and the power will be controlled.

The CCG current control using MPC-WM diagram is depicted in Fig. 7.14 and its basic code is presented in Algorithm 7.3.

7.3.3 Simulation results

This section introduces the simulation results using the data presented in Table B.1. The horizons are $n_u = 2$, $n_y = 3$, $W_{yy,g} = 10^4 I$, and $W_{uu,g} = 1.5 \cdot 10^4 I$.

The behavior of the grid current during the step tests for grid power control, in which the grid active power (P_g) varies from 2000 to 1000 W and the grid reactive power (Q_g) varies from −500 to 1000 var, is depicted in Figs. 7.15 and 7.16. It can be noticed that the MPC-WM controller reaches the references of the grid current elements, and the THD of the current is 5.51%.

7.3.4 Experimental results

This section introduces the experimental results obtained in the setup by using the data presented in Table B.1. The horizons are $n_u = 2$, $n_y = 3$, $W_{yy,g} = 10^4 I$, and $W_{uu,g} = 1.5 \cdot 10^4 I$.

The behavior of grid power control in the step tests, in which the grid active power (P_g) varies from 500 to 750 W and the grid reactive power (Q_g) varies from 0 to −300 var, is depicted in Figs. 7.17(a) and 7.17(b). The settling time is about 1.5 ms, as can be demonstrated in Fig. 7.18(a), and the Q_g remains constant at 0 var, as depicted in Fig. 7.18(b).

In this way, it can be noticed that the MPC-WM calculates the grid voltage vector necessary for the grid current vector components and power to reach their references. The actions of the grid current vector elements are depicted in Figs. 7.19(a)–(b). The grid current reference vector is calculated by means of

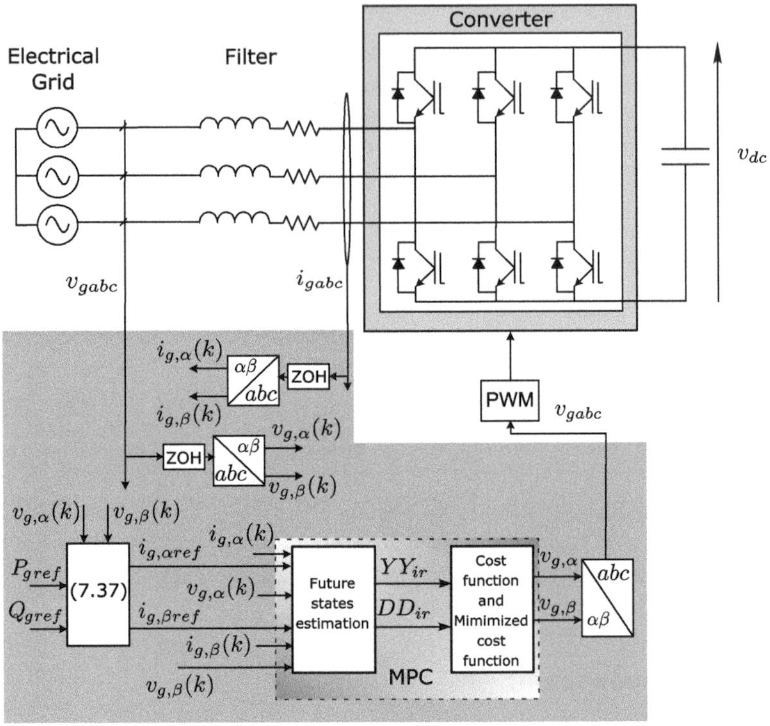

FIGURE 7.14 MPC-WM. CCG current control diagram.

Algorithm 7.3 MPC-WM for CCG current control algorithm

---

while MPC-WM - grid current control active **do**
 $\vec{i}_{g,\alpha\beta}(k)$, $\vec{v}_{g,\alpha\beta}(k)$, $\vec{i}_{g,\alpha\beta ref}(k) \leftarrow$ % System measurements
 Calculate $A_{ad,g}$, $B_{ad,g}$
 Calculate AA_g, BB_g, function of $(A_{ad,g}, B_{ad,g}, n_y, n_u)$
 $YY_{ref,g} \leftarrow$ Assign values of $\vec{v}_{r,\alpha\beta ref}$
 $W_{yy,g}$, $W_{uu,g} \leftarrow$ Assign values
 Calculate VV_g by using Eq. (7.38)
 Apply $\vec{v}_{g,\alpha\beta}$ to the PWM technique.
end while

---

the power presented in Eq. (7.37). The oscillations occur due to the switching of the inverter.

The grid current and voltage during the reactive and active grid power step test are depicted in Figs. 7.20 and 7.21, respectively. It can be noticed that the grid voltage remains constant, the grid current increases its amplitude when the active power increases its value, and the grid current changes its angle when the reactive power changes its values in accordance with the model representation.

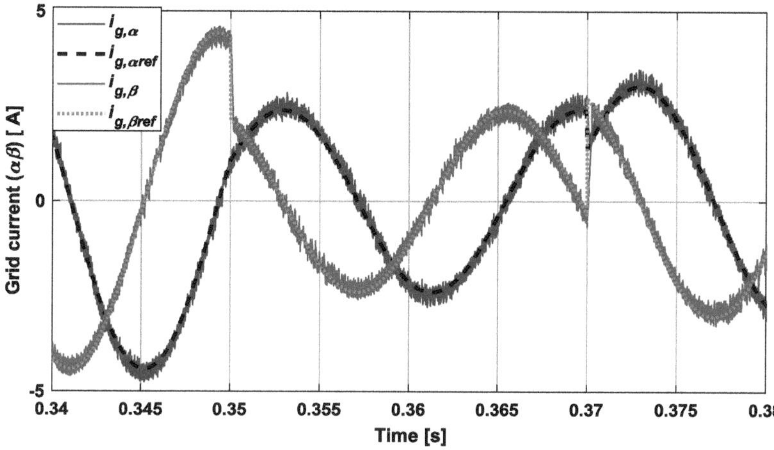

FIGURE 7.15 MPC-WM. Simulation result of the grid current behavior during the power step tests.

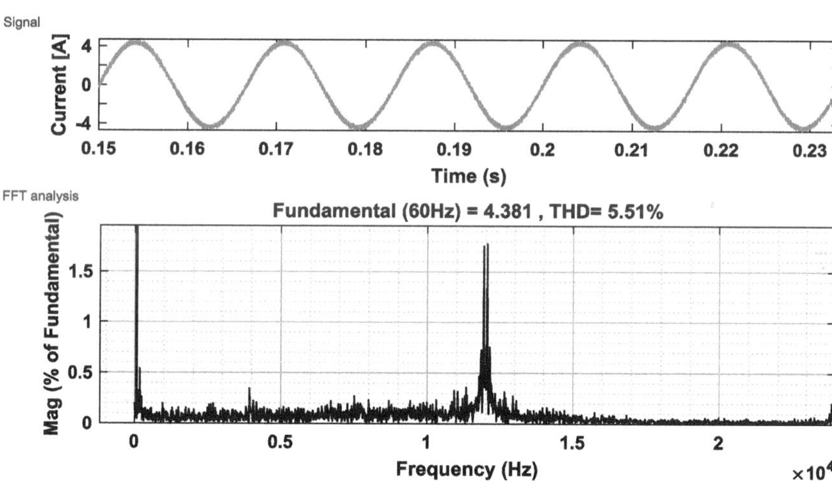

FIGURE 7.16 MPC-WM: THD of the grid current.

7.4 Information about the choice of weighting matrices and horizons values

This section presents the guidelines to the choice of weighting matrices and horizon values. The main information to reach a satisfactory behavior of the output by using MPC-WM is based on the information presented in [4] and the presented tests in this chapter.

Regarding the horizons, we take $n_y = n_u +$plant settling time. When $n_u = 1$, the behavior often is near to being open-loop. The control action magnitude

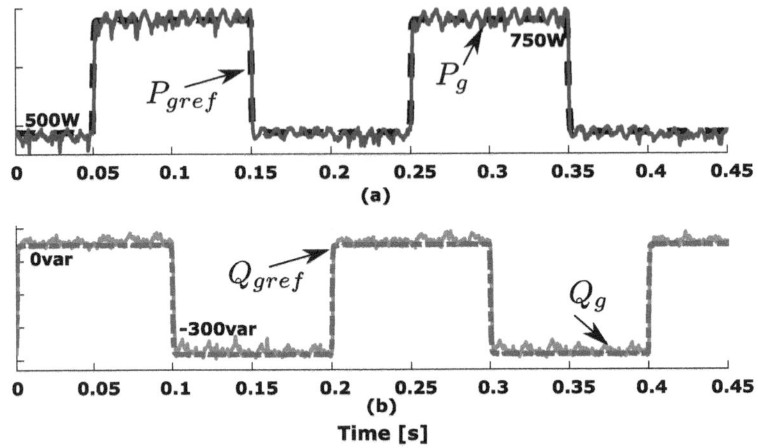

FIGURE 7.17 MPC-WM. Step tests for grid power control: (a) active power, (b) reactive power.

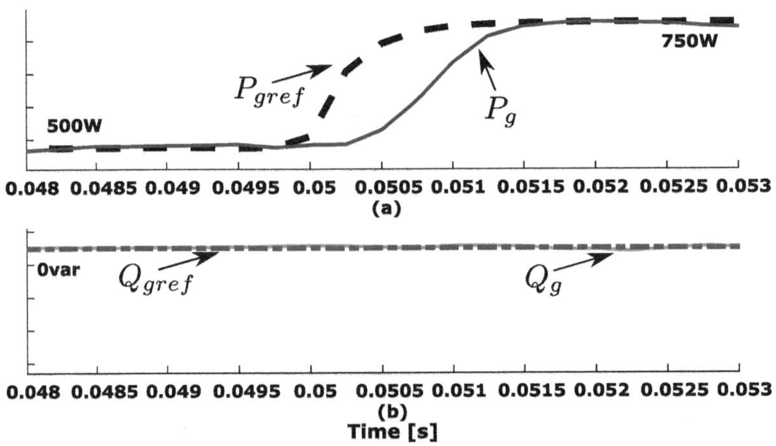

FIGURE 7.18 MPC-WM. Detail of grid active power step: (a) active power, (b) reactive power.

represents average values and the outputs reach the references. When $n_u \geq 3$, the "global optimum" can be achieved. Large values of n_y and n_u increase the computational cost, and some assumptions will not be true as considered in this chapter.

Regarding the choice of the weight matrices' values, we can choose them by means of the analysis of the cost functions in Eqs. (7.12), (7.25), and (7.36). The overshoot can be decreased when the values of the elements of $W_{uu,ir}$, $W_{uu,pq}$, and $W_{uu,g}$ are greater than zero, in other words, the actuator of the system has infinite energy.

The elements of $W_{yy,ir}$, $W_{yy,pq}$, and $W_{yy,g}$ can improve the settling time due to the emphasis on each predicted output. Sometimes when the values of

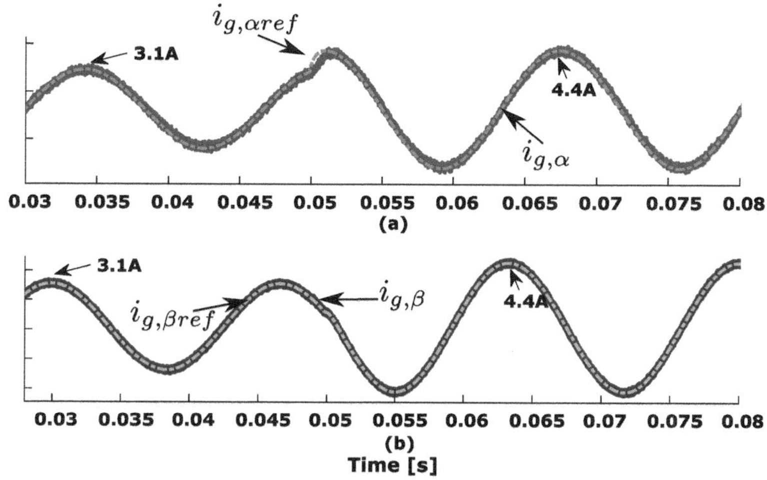

FIGURE 7.19 MPC-WM. Grid current during power step tests: (a) real component of the grid current vector, (b) imaginary component of the grid current vector.

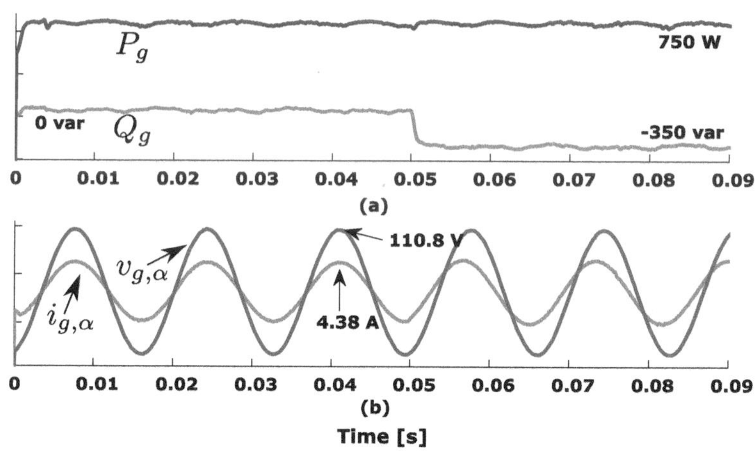

FIGURE 7.20 MPC-WM. Grid current during the reactive power step: (a) active and reactive power, (b) grid voltage and current.

the matrices have a large change they may have a negligible effect. Due to simplicity, in this book the weight matrices' values are a constant plus the identity matrix.

7.5 Summary

This chapter describes the rotor current control for DFIG, the DPC for DFIG, and the grid current control for CCG using the MPC-WM technique in normal

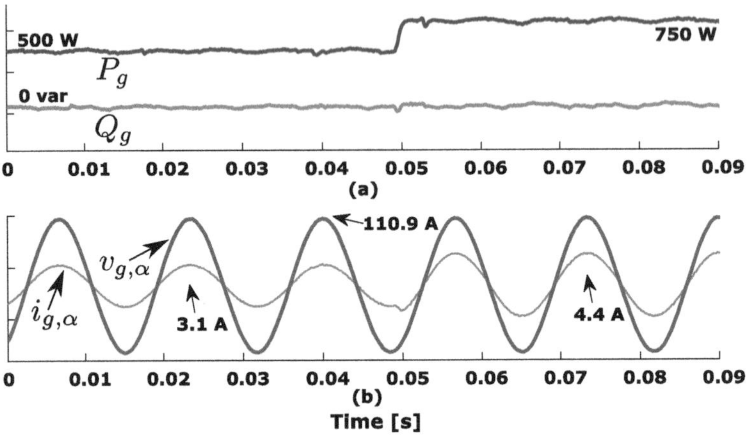

FIGURE 7.21 MPC-WM. Grid current during the active power step: (a) active and reactive power, (b) grid voltage and current.

operation conditions. The MPC-WM employs the discrete representation of the mentioned systems. In this way, the DFIG is represented in the dq frame and the CCG is expressed in the $\alpha\beta$ frame. In both cases, the MPC-WM allows reaching the references and in accordance with the presented experimental results. Additionally, information about the weighting matrices' choice is described, and this technique operates using a constant switching frequency due to the PWM. In this way, this scheme may be of interest for the DFIG and CCG current control applications.

Chapter 8

Fundamentals of the model predictive repetitive control

Alfeu J. Sguarezi Filho
Federal University of ABC, Santo André, SP, Brazil

8.1 Fundamentals of repetitive control

The repetitive control (RC) is a technique that employs the internal model principle (IMP) [108,109]. In this principle the signal model has to exist in the control system. Hence, it will be possible to follow or to reject this signal.

8.1.1 IMP for any periodic signal

A periodic signal that has (T_{pp}) period can be represented by using the Fourier series in exponential form written as [108,110]

$$r_{rp}(t) = \sum_{nn=-\infty}^{\infty} a_{nn} e^{j\frac{2\cdot\pi\cdot nn\cdot t}{T_{pp}}}. \tag{8.1}$$

For the inclusion of Eq. (8.1) into the control system loop by using the IMP, one can use its transfer function in Laplace domain which can be represented as [110]

$$R_{rp}(s) = \frac{1}{s}\prod_{nn=1}^{\infty}\frac{\left(\frac{2\pi nn}{T_{pp}}\right)^2}{\left(\frac{2\pi nn}{T_{pp}}\right)^2 + s^2} = \frac{T_{pp}e^{\frac{-sT_{pp}}{2}}}{1-e^{-sT_{pp}}}. \tag{8.2}$$

However, $T_{pp}e^{\frac{-sT_{pp}}{2}}$ represents a delay term with gain T_{pp}. In this case, the transfer function $\frac{1}{1-e^{-sT_{pp}}}$ is enough to insert into the closed loop. The implementation of this transfer function can be achieved by means of a positive feedback loop by using $e^{-sT_{pp}}$ as depicted in Fig. 8.1. It can be noticed that this model has poles on the imaginary axis, $s = \frac{jnn}{T_{pp}}$, $nn \in \aleph$, and the model has infinite gain at frequencies $\frac{nn}{T_{pp}}$, $\forall\, nn \in \aleph$, regarding the frequencies point of view.

https://doi.org/10.1016/B978-0-32-390964-8.00017-8

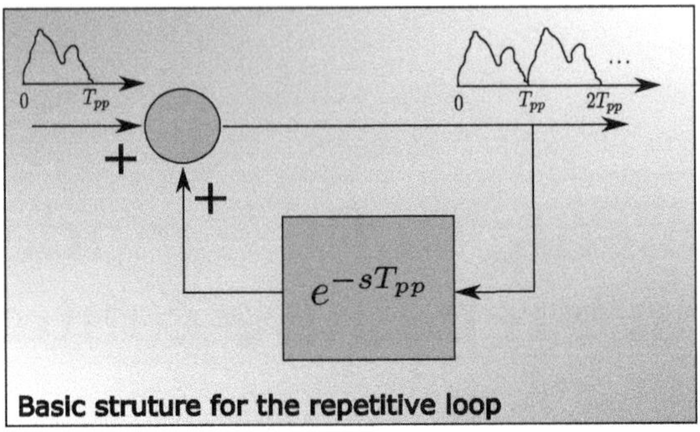

FIGURE 8.1 Control system loop by using the IMP.

The implementation of the IMP presented in the transfer function of Eq. (8.2) in the discrete-time domain can be written as

$$G_{rr}(z) = \frac{z^{-N}}{1 - z^{-N}} = \frac{1}{z^{-N} - 1}, \tag{8.3}$$

where we have the relationship $N = T_{pp}/T$ with T being the sampling time.

One controller that can realize the IMP is the PID (proportional-integral-derivative) which can be represented in the Laplace domain as

$$PID = k_p + \frac{k_i}{s} + k_d s. \tag{8.4}$$

In the case of Eq. (8.4), the integrator $\frac{1}{s}$ is an internal model of DC signals that allows null steady state error. The proportional and derivative terms increase the controller performance and robustness.

8.1.2 Basic RC structure and design

The RC can be inserted in a plug-in scheme. In this way, the RC is inserted in the control loop where there is an existing nominal control $G_{crp}(z)$ and the system plant $G_{prp}(z)$ as depicted in Fig. 8.2. This controller allows rejecting the disturbance $D_{rp}(z)$ across a broad frequency spectrum and stabilizing the plant $G_{prp}(z)$. Additionally, in the RC formulation, we introduce a filter $H_{rp}(z)$ and $G_{rp}(z)$ to increase the robustness and closed-loop stability, respectively. The term $H_{rp}(z)$ is in the positive feedback connection. This filter can be a low-pass null-phase or any finite impulse filter with similar behavior.

The condition for the stability can be found by means of Eq. (8.5). The $G_{crp}(z)$ design can be made in accordance with the stability criterion in the context of the small gain theorem as presented in Eqs. (8.5)–(8.7) [110]:

$$\left|\left(1 - \frac{G_{crp}(z)G_{prp}(z)G_{rp}(z)}{1 + G_{crp}(z)G_{prp}(z)} H(z)\right)\right|_{z=e^{j\omega T}} < 1, \quad \forall \omega \in \left[0, \frac{\pi}{T}\right], \tag{8.5}$$

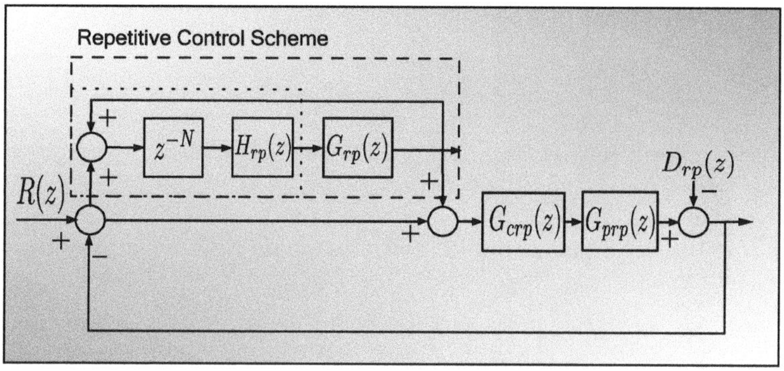

FIGURE 8.2 RC in a plug-in scheme.

$$\left|H_{rp}(z)\right|_{z=e^{j\omega T}} < 1, \quad \forall\, \omega \in \left[0, \frac{\pi}{T}\right], \tag{8.6}$$

$$\left|1 - \frac{G_{crp}(z)G_{prp}(z)}{1 + G_{crp}(z)G_{prp}(z)}\right|_{z=e^{j\omega T}} < 1, \quad \forall\, \omega \in \left[0, \frac{\pi}{T}\right]. \tag{8.7}$$

8.2 Fundamentals of model predictive repetitive control

This section introduces the concepts of the model predictive repetitive control (MPRC).

8.2.1 Periodic signals representation

Another way to represent the periodic reference signals was developed in [111, 112] by using the discrete Fourier transform (DFT). In this way, the periodic reference signal has the period T_{pp} and it is uniformly sampled to reach the discrete sequence $r_{rp}(k)$, $k = 0, 1, \ldots, M_x - 1$ with interval Δt and $M_x = \frac{T}{\Delta t}$. It is assumed that there are no frequency components greater than $1/(2\Delta t)$. This signal can be represented using the DFT and written as

$$r_{rp}(k) = \frac{1}{M_x} \sum_{nn=0}^{M_x-1} R\left(e^{j\frac{2 \cdot \pi \cdot nn}{M_x}} \right) e^{j\frac{2 \cdot k \cdot \pi}{M_x}}, \tag{8.8}$$

where the frequency components in the signal are represented as $R\left(e^{j\frac{2\cdot\pi\cdot nn}{M_x}}\right)$, $nn = 0, 1, 2, \ldots, M_x - 1$. The representation of Eq. (8.8) by using the z-transform can be performed as

$$R_{rpm}(z) = \sum_{k=0}^{M_x-1} r_{rp}(k) z^{-k}, \tag{8.9}$$

where the fundamental frequency is $\omega = \frac{2\pi}{M_x}$. Hence, Eq. (8.9) can be rewritten using Eq. (8.8) as

$$R_{rpm}(z) = \sum_{l=-(M_x-1)/2}^{(M_x-1)/2} R\left(e^{j\omega \cdot l}\right) H_l(z), \tag{8.10}$$

where $H_l(z)$ is described as the lth order harmonic filter which can be represented as

$$H_l(z) = \frac{1}{M_x}\frac{-z^{-M_x}+1}{1-e^{j\omega l}z^{-1}} = \frac{1}{M_x}\left(e^{j\omega \cdot l}z^{-1} + \cdots + e^{j\omega \cdot l(M_x-1)}z^{-(M_x-1)}\right), \tag{8.11}$$

where the inclusion of the zero frequency can be reached when the number M_x is odd.

Additionally, $H_l(z)$ in Eq. (8.11) can be managed as two-cascaded filters as in [109]:

$$H_l(z) = \underbrace{\left(\frac{1}{M_x}\frac{1}{1-e^{j\omega \cdot l}z^{-1}}\right)}_{\text{generalized integrator at } \omega \cdot l} \cdot \overbrace{\left(1-z^{-M_x}\right)}^{\text{combined filter with notches at all harmonic frequencies}}. \tag{8.12}$$

At this moment, Eq. (8.10) can be represented in terms of imaginary ($\mathcal{I}m$) and real ($\mathcal{R}e$) elements of the frequency components $R\left(e^{j\omega \cdot l}\right)$ as follows:

$$\begin{aligned} R_{rpm}(z) = & \left(\frac{1}{M_x}\frac{z^{-M_x}}{1-z^{-1}}\right) R\left(e^{j0}\right) \\ & + \sum_{l=1}^{(M_x-1)/2} \left[\mathcal{I}m\left(R\left(e^{j\omega \cdot l}\right)\right) F_{xi}(z) + \mathcal{R}e\left(R\left(e^{j\omega \cdot l}\right)\right) F_{xr}(z)\right], \end{aligned} \tag{8.13}$$

where the $F_{xr}(z)$ and $F_{xi}(z)$ are the lth-order harmonic filter as follows:

$$F_{xi}(z) = \frac{2\sin(\omega \cdot l)z^{-1}(1-z^{-M_x})}{M_x(1-2\cos(\omega \cdot l)z^{-1}+z^{-2})} \tag{8.14}$$

and

$$F_{xr}(z) = \frac{2(1-\cos(\omega \cdot l)z^{-1}(1-z^{-M_x}))}{M_x(1-2\cos(\omega \cdot l)z^{-1}+z^{-2})}. \tag{8.15}$$

It can be noticed that the denominator of Eqs. (8.14) and (8.15) is the representation of a resonant controller [109].

8.2.1.1 *Signal generator for the MPRC*

The z-transform of the reference signal $R_{rpm}(z)$ presented in Eq. (8.13) allows obtaining the main frequency components in a periodic signal. It can be noticed that the integrator factor $(1 - z^{-1})$ employed in feedback control systems appears in its denominator and on the left-hand side of Eq. (8.12). This item at zero frequency has the function to eliminate unknown system bias or reject constant disturbances in the reference signal. Additionally, the introduction of the factor $(1 - 2\cos(\omega l)z^{-1} + z^{-2})$ can be seen in Eqs. (8.14) and (8.15) for an arbitrary frequency. In this way, $D(z)$ is the signal polynomial to be inserted in the repetitive controller which can be represented as [112]:

$$\begin{aligned} D(z) &= (1 - z^{-1}) \prod_{l_1}^{l=l_n} (1 - 2\cos(\omega \cdot l)z^{-1} + z^{-2}) \\ &= 1 + d_1 z^{-1} + d_2 z^{-2} + \cdots + d_\rho z^{-\rho}, \end{aligned} \tag{8.16}$$

where ρ is the order of the filter and l denotes the harmonic components.

When the system is multiple input–multiple output (MIMO), the individual references signals are enough. In this way, $D(z)$ can be represented as per the common denominator of the resulting polynomials.

8.2.2 MPRC technique

As presented in Chapter 4, the MPC predicts the system future states by using its mathematical model. Then, the predictive controller calculates the input to reach the references by using the future system actions and the minimized cost function that can have constraints in another parameters.

The MPRC was developed in [112–114]. This technique adds the repetitive controller behavior in the classical MPC in which the disturbance rejection capability of the controller is increased.

The MPRC can be designed using the expanded plant model (in state space representation) due to the introduction of the required polynomial $D(z)$. In this way, the MPC will be implemented by means of this expanded state space model of the plant. The space state of the model can be represented in Eq. (4.11), where the matrices' and vectors' sizes are known and the term $G_{gd}d_d(k)$ represents a perturbation in this model. In this way, each entry of the mentioned term $(G_{gd}d_d(k))$ can be written as the inverse z-transform of $1/D(z)$. When the system operates in steady state, $(G_{gd}d_d(k))$, for every $k \geq k_0$, it can be represented as per the difference equation as follows:

$$D(q^{-1})(G_{gd}d_d(k)) = 0, \tag{8.17}$$

where q^{-1} represents the backward shift operator and k_0 is the initial sampling instant.

The expanded state space model can be achieved by multiplying Eq. (4.11), $x_x(k+1) = A_{ad}x_x(k) + B_{bd}u_u(k) + G_{gd}d_d(k)$ and $y_y(k+1) = C_{cd}x_x(k+1)$, by $D(q^{-1})$ and it is given by

$$\begin{aligned} x_{xx}(k+1) &= A_{ad}x_{xx}(k) + B_{bd}u_{xx}(k), \\ D(q^{-1})y_y(k+1) &= C_{cd}x_{xx}(k+1), \end{aligned} \tag{8.18}$$

where the subscript xx denote the auxiliary variables.

These variables can be represented by means of the multiplication of the polynomial $D(q^{-1})$ with the input $u_u(k)$ and the state $x_x(k)$, as it can be expressed in

$$x_{xx}(k) = D(q^{-1})x_x(k), \quad u_{xx}(k) = D(q^{-1})u_u(k). \tag{8.19}$$

Substituting $x_{xx}(k+1)$ into the output equation and applying the polynomial by means of the backward shift operator over the mentioned equation we can obtain

$$\begin{aligned} y_y(k+1) = &-d_1y_y(k) - d_2y_y(k-1) - \cdots \\ &- d_\rho y_y(k-\rho+1) + C_{cd}A_{ad}x_{xx}(k) + C_{cd}B_{bd}u_{xx}(k). \end{aligned} \tag{8.20}$$

The state space representation in discrete form of the expanded model of the system as per the introduction of the polynomial can be presented in

$$\begin{aligned} x_{ee}(k+1) &= A_{ee}x_{ee}(k) + B_{ee}u_{xx}(k), \\ y_y(k) &= C_{ee}x_{ee}(k), \end{aligned} \tag{8.21}$$

where ee is related to the expanded matrices,

$$x_{ee}(k) = \begin{bmatrix} x_{xx}(k)^T & y_y(k)^T & \ldots & y_y(k-\rho+1)^T \end{bmatrix}^T, \tag{8.22}$$

$$A_{ee} = \begin{bmatrix} A_{ad} & O \\ \hat{M} & A_{pp} \end{bmatrix} \hat{M} = \begin{bmatrix} C_{cd}A_{ad} \\ O \end{bmatrix}, \tag{8.23}$$

$$A_{pp} = \begin{bmatrix} -d_1I & -d_2I & \ldots & -d_{\rho-1}I & -d_\rho I \\ I & O & O & \ldots & O \\ 0 & I & O & \ldots & O \\ \vdots & \ddots & \ddots & \ddots & \vdots \\ O & O & \ldots & I & O \end{bmatrix}, \tag{8.24}$$

$$B_{ee} = \begin{bmatrix} B_{bd} \\ C_{cd}B_{bd} \\ O \\ \vdots \\ O \end{bmatrix} C_{ee} = \begin{bmatrix} O & I & O & \ldots & O \end{bmatrix}, \tag{8.25}$$

and I and O indicate identity and zero matrix, respectively.

Eq. (8.26) expresses the actions of the future plant outputs by means of the expanded state model in discrete form, namely

$$\mathbf{X_{pr}} = A_{pr}x_{ee}(k_i) + B_{pr}\mathbf{U_{pr}}, \tag{8.26}$$

where each of the terms can be described as follows:

$$\mathbf{X_{pr}} = \left[x_{ee}(k_i+1|k_i)^T \quad \ldots \quad x_{ee}(k_i+n_y|k_i)^T\right]^T, \tag{8.27}$$

$$\mathbf{U_{pr}} = \left[u_{xx}(k_i)^T \quad \ldots \quad u_{xx}(k_i+n_u-1)^T\right]^T, \tag{8.28}$$

$$A_{pr} = \begin{bmatrix} A_{ee} \\ A_{ee}^2 \\ \vdots \\ A_{ee}^{n_p} \end{bmatrix} B_{pr} = \begin{bmatrix} B_{ee} & O & \ldots & O \\ A_{ee}B_{ee} & B_{ee} & \ldots & O \\ A_{ee}^2 B_{ee} & A_{ee}B_{ee} & \ldots & O \\ \vdots & \ddots & \ddots & \vdots \\ A_{ee}^{n_y-1}B_{ee} & A_{ee}^{n_y-2}B_{ee} & \ldots & A_{ee}^{n_y-n_u}B_{ee} \end{bmatrix}. \tag{8.29}$$

In Eq. (8.26), there is a new state space formulation in which the sampling instant k_i always satisfies $k_i > 0$. The measurement of the state provides information about the system. Also $\mathbf{X_{pr}}$ and $\mathbf{U_{pr}}$ are the predicted states and control values for the state and input, respectively. It is assumed that for $n_u \leq n_y$ and for $k \geq n_u$, all the control samples $u_{ee}(k_i+k)$ turn null to reduce the computational cost.

To reach the predicted control values (U_{pr}) by means of MPRC, the cost function is presented as

$$J = \mathbf{X_{pr}}^T \overline{Q_i} \mathbf{X_{pr}} + \mathbf{U_{pr}}^T \overline{R_v} \mathbf{U_{pr}}, \tag{8.30}$$

where $\overline{R_v}$ expresses a block diagonal matrix that has the $R_v I$ form, in which R_v represents a symmetric positive definite matrix and $\overline{Q_i}$ expresses a block diagonal matrix that has the $Q_i I$ form, in which Q_i represents a symmetric positive semidefinite matrix. Additionally, it is possible to add constraints as presented in Section 4.3.1. So, the minimized cost function in Eq. (8.30) can be rewritten as presented by

$$\begin{aligned} J = \mathbf{U_{pr}}^T (B_{pr}^T \overline{Q_i} B_{pr} + \overline{R_v})\mathbf{U_{pr}} + 2\mathbf{U_{pr}}^T B_{pr}^T \overline{Q_i} A_{pr} x_{ee}(k_i) \\ + x_{ee}(k_i)^T A_{pr}^T Q_i A_{pr} x_{ee}(k_i). \end{aligned} \tag{8.31}$$

When there are no constraints in the optimal vector $\mathbf{U_{pr}}$, it can be represented by means of Eq. (8.31) as

$$\mathbf{U_{pr}} = -(B_{pr}^T \overline{Q_i} B_{pr} + \overline{R_v})^{-1} B_{pr}^T \overline{Q_i} A_{pr} x_{ee}(k_i). \tag{8.32}$$

Extracting the values of u_{ee} from the vector $\mathbf{U_{pr}}$ and as per Eq. (8.19), the input control signal to be inserted in the k_i sampling time is presented by

$$u_u(k_i) = u_{xx}(k_i) - d_1 u_u(k_i - 1) - \cdots - d_\rho u_u(k_i - \rho). \tag{8.33}$$

When the MPRC system operates using a high sampling rate, this permits obtaining the past sample with minimal error. In this case, the mentioned frequency is compared to the fundamental frequency of the plant of the system. In this context, the MPRC will be employed to track a reference signal. Hence, the insertion of the reference signal into the expanded state space model is mandatory.

In this context, the augmented output vector $x_{ee}(k_i)$, after inserting the reference signal, can be written in Eq. (8.22) considering the error signal $\vec{e}(k)$. Hence, there is a new equation to represent $x_{ee}(k_i)$, namely

$$x_{ee}(k) = \begin{bmatrix} x_{xx}(k)^T & \vec{e}(k)^T & \ldots & \vec{e}(k - \rho + 1)^T \end{bmatrix}^T, \tag{8.34}$$

where $\vec{e}(k)$ represents the error signal which can be represented as in Eq. (8.35). And, it is the difference of the output signal $(y(k_i))$ and the reference signal $(r(k_i))$, that is,

$$\vec{e}(k) = y(k_i) - r(k_i). \tag{8.35}$$

In this book, the CCG and DFIG are controlled using voltage vectors that are computed at each sampling time by the MPRC. When the voltage vector is computed by means of MPRC, Eq. (8.33) allows accomplishing this task. Hence, the voltage vector components will be the input of the PWM algorithm of the three-phase power CCG or of the converter connected to the rotor of the DFIG. The MPRC can be described applying the scheme presented in Fig. 8.3, and its basic code can be seen in Algorithm 8.1. In this book, when the MPRC employs the PWM, it is called MPRC-WM.

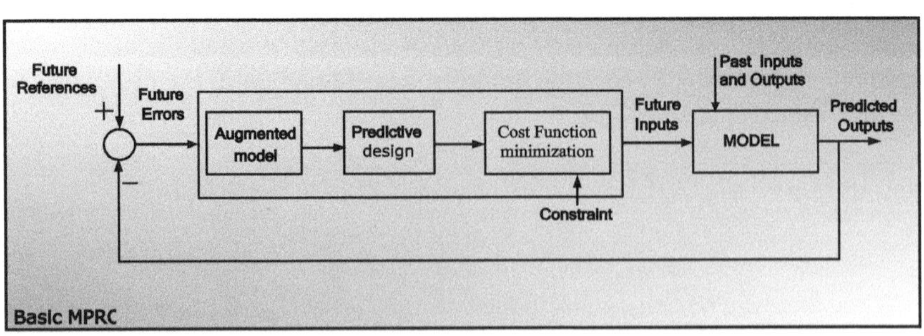

FIGURE 8.3 MPRC diagram.

Algorithm 8.1 MPRC algorithm

while Controller Active **do**
  $output, e(k) \leftarrow$ System measurements
  $[x_{xx}(k), \quad e] = \text{func}_1(x_x(k), \vec{e}(k), D(z), x_{xx})$
  Compute: $[A_{ad}, \quad B_{bd}, \quad C_{cd}]$
  $[A_{ee}, \quad B_{ee}] = \text{func}_2(A_{ad}, B_{bd}, C_{cd}, D(z))$
  $[A_{pr}, \quad B_{pr}] = \text{func}_3(A_{ee}, B_{ee}, n_u, n_u)$
  $\overline{Q_i}, \overline{R_v} \leftarrow$ Assign values
  $x_{ee}(k) = \text{func}_4(x_{xx}(k))$
  $\mathbf{U_{pr}} = (x_{ee}(k), A_{pr}, B_{pr}, \overline{Q_i}, \overline{R_v})$
  $input(k) = \text{func}_5(\mathbf{U_{pr}}, D(z), input)$
end while

8.3 Summary

This chapter describes the fundamentals of the MPRC. This MPC is a group of controllers that predict the system action and compute the input using the minimized cost function at each sampling time. In this type of scheme (MPRC), the repetitive technique actions are inserted into MPC. Hence, one can reach the references and reject the disturbances. The concepts introduced are fundamental to the repetitive control and MPRC. In this way, the described predictive techniques permit designing MPRCs to DFIG and CCG.

Chapter 9

MPRC-WM for DFIG and three-phase CCG operation under voltage distortions

Alfeu J. Sguarezi Filho, Angelo S. Lunardi, and Eliomar R. Conde D.
Federal University of ABC, Santo André, SP, Brazil

9.1 Representation of voltage distortions

The voltage distortions in power systems are the voltage of fundamental frequency plus harmonic components which are caused by the nonlinear loads or voltage sources. In this context, the mentioned harmonic components can increase the losses and magnetic saturation in the generator or transformers. In normal operation, the grid voltage has total harmonic distortion (THD) lower than 5% [115]. Hence, in this chapter an objective is to inject a stator or grid current with low THD by using MPRC.

The representation of the symmetrical grid voltage for any phase under steady state condition with harmonic components can be performed by means of Eq. (8.1), and after some algebraic manipulation it can be described as follows:

$$v(t) = v_{l_1} + \sum_{l=l_2}^{\infty} v_l, \tag{9.1}$$

where v_{l_1} expresses the fundamental frequency of the electrical grid voltage and l represents the harmonic components. The THD is the relationship between the root mean square (rms) value of the fundamental frequency element and the sum of the rms values of the components greater than twice the distorted voltage, what can be computed by means of the IEEE Standard 519-1992 [116,117]:

$$\%THD = \frac{\sqrt{\sum_{l=l_2}^{\infty} v_{l-rms}}}{v_1} \cdot 100, \tag{9.2}$$

where $v_{l-rms} = \sqrt{\frac{1}{T_{pp}} \int_0^{T_{pp}} v_l^2 \cdot dt}$ represents the rms value of each harmonic component v_l and v_{1-rms} expresses the rms value of the fundamental frequency v_{l_1}. Normally, we consider harmonic components up to the 50th order. However,

Model Predictive Control for Doubly-Fed Induction Generators and Three-Phase Power Converters
https://doi.org/10.1016/B978-0-32-390964-8.00018-X

harmonics greater than the mentioned value can be inserted. Also, the THD can be applied to the current. In this context, it is just a change of the voltage (v) representation by current (i) notation in Eqs. (9.1) and (9.2).

9.2 Model of DFIG under stator distorted voltage

This section introduces the model of the DFIG when the distorted voltages at the stator occur. In this way, we will employ the rotor model as presented in Eq. (3.21), namely $\vec{v}_{r,dq} = R_r \vec{i}_{r,dq} + \frac{d}{dt}\left(L_m \vec{i}_{s,dq} + L_r \vec{i}_{r,dq}\right) + j(\omega_s - \omega_r)\left(L_m \vec{i}_{s,dq} + L_r \vec{i}_{r,dq}\right)$.

The rotor voltage vector is represented by means of the rotor current vector and the stator flux vectors elements. The representation is achieved by substituting the stator current in Eq. (2.50) ($\vec{\lambda}_{s,dq} = L_s \vec{i}_{s,dq} + L_m \vec{i}_{r,dq}$) into Eq. (3.21). Hence, the rotor voltage vector becomes

$$\vec{v}_{r,dq} = (R_r + j\sigma L_r \omega_{sl})\,\vec{i}_{r,dq} + \sigma L_r \frac{d\vec{i}_{r,dq}}{dt} + \frac{L_m}{L_s}\left(\frac{d}{dt} + j\omega_{sl}\right)\vec{\lambda}_{s,dq}, \tag{9.3}$$

which means

$$v_{r,d} = R_r i_{r,d} - \omega_{sl}\sigma L_r i_{r,q} + \sigma L_r \frac{di_{r,d}}{dt} + \frac{L_m}{L_s}\left(\frac{d\lambda_{s,d}}{dt} - \omega_{sl}\lambda_{s,q}\right) \tag{9.4}$$

and

$$v_{r,q} = R_r i_{r,q} + \omega_{sl}\sigma L_r i_{r,d} + \sigma L_r \frac{di_{r,q}}{dt} + \frac{L_m}{L_s}\left(\omega_{sl}\lambda_{s,d} + \frac{d\lambda_{s,q}}{dt}\right). \tag{9.5}$$

The rate $\frac{d\vec{\lambda}_{s,dq}}{dt}$ can be represented as per Eq. (2.46) ($\vec{v}_{s,dq} = \vec{i}_{s,dq} R_s + \frac{d\vec{\lambda}_{s,dq}}{dt} + j\omega_s \vec{\lambda}_{s,dq}$) when the stator resistance is neglected and it can be written as follows:

$$\begin{aligned} \vec{v}_{s,dq} - j\omega_s \vec{\lambda}_{s,dq} &= \frac{d\vec{\lambda}_{s,dq}}{dt}, \\ \vec{\lambda}_{s,dq} &= \frac{\vec{v}_{s,dq}}{\frac{d}{dt} + j\omega_s}. \end{aligned} \tag{9.6}$$

The DFIG model considering the effects of the stator voltage can be reached by substituting the stator flux vector of Eq. (9.6) into Eq. (9.3). Hence, the new expression can be written as follows:

$$\begin{aligned} \vec{v}_{r,dq} = {} & (R_r + j\sigma L_r \omega_{sl})\,\vec{i}_{r,dq} \\ & + \sigma L_r \frac{d\vec{i}_{r,dq}}{dt} + \frac{L_m}{L_s}\left(\frac{d}{dt} + j\omega_{sl}\right)\left(\frac{\vec{v}_{s,dq}}{\frac{d}{dt} + j\omega_s}\right). \end{aligned} \tag{9.7}$$

The representation of Eq. (9.7) in Laplace domain by means of the complex transfer function concept [21] can be written as:

$$\vec{v}_{r,dq}(s) = (R_r + j\sigma L_r\omega_{sl})\,\vec{i}_{r,dq}(s) + \sigma L_r\,\vec{i}_{r,dq}(s)\cdot s + \frac{L_m}{L_s}(s + j\omega_{sl})\left(\frac{\vec{v}_{s,dq}}{s + j\omega_s}\right). \tag{9.8}$$

It can be noticed in Eqs. (9.7) and (9.8) that the stator voltage impacts the rotor voltage equations even in the event of voltage distortions. The works [118, 119] studied the control and the DFIG model in this situation.

9.2.1 Influence of distorted voltage in the stator active and reactive power representation

The voltages in the presence of nonsinusoidal conditions in a three-phase system can be represented as:

$$\begin{aligned} v_{al} &= V_l \sin(l\omega + \theta_l), \\ v_{bl} &= V_l \sin\left(l\omega + \frac{2l\pi}{3} + \theta_l\right), \\ v_{cl} &= V_l \sin\left(l\omega - \frac{2l\pi}{3} + \theta_l\right). \end{aligned} \tag{9.9}$$

This representation is also valid for the currents. When the system operates under balanced conditions, the lth harmonic (frequency $= l\omega$) phase c lags $l \times 240°$ and the phase b lags $l \times 120°$ behind that of the same harmonic in phase a. When the triple harmonics occur, the vector is cophasial due to the shifting of three times 240° or three times 120°. Hence, they are zero sequence phasors. The sequences can be written as [120]:

- Positive sequence, harmonic elements of the order $3 \cdot l + 1$;
- Zero sequence, harmonic elements of the order $3 \cdot l$;
- Negative sequence, harmonic elements of the order $3 \cdot l + 2$.

The system reference transformations can be useful in this case due to the reduction in the number of equations. In this way, to illustrate this situation, the voltage vector $\vec{v}$ with harmonic components can be represented as:

$$\vec{v}_{dq} = \underbrace{\vec{v}_{dq-l_1}}_{\text{fundamental frequency}} + \underbrace{\vec{v}_{dq-l_5} + \vec{v}_{dq-l_7}}_{\text{harmonic components}}. \tag{9.10}$$

Each harmonic component of the voltage vector in Eq. (9.10) is in the same dq frame, considering the fundamental frequency $dq - l_1$, $dq - l_5 =$ 5th and $dq - l_7 =$ 7th harmonic components, respectively, and it can be described using an arbitrary vector $\vec{v}_{\alpha\beta}$ as presented in [118,119]:

$$\begin{aligned}
\vec{v}_{dq-l_1} &= \vec{v}_{\alpha\beta} e^{-j\omega_s t}, \\
\vec{v}_{dq-l_5} &= \vec{v}_{l_5} e^{-j6\omega_s t}, \\
\vec{v}_{dq-l_7} &= \vec{v}_{l_7} e^{j6\omega_s t}, \\
\vec{v}_{l_5} &= \vec{v}_{\alpha\beta} e^{j5\omega_s t}, \\
\vec{v}_{l_7} &= \vec{v}_{\alpha\beta} e^{-j7\omega_s t}.
\end{aligned} \tag{9.11}$$

Eq. (9.11) also employs the respective quantities of the each harmonic component instead of the arbitrary vector $\vec{v}_{\alpha\beta}$. Additionally, the presented relationships are also valid for the current vector.

When the stator voltage in the dq frame is under the presence of the harmonic components, it can be described as follows:

$$\vec{v}_{s,dq} = \sum_{l=l_1}^{l_n} \vec{v}_{s,dq-l_n} = \vec{v}_{s,dq-l_1} + \vec{v}_{s,dq-l_2} + \cdots + \vec{v}_{s,dq-l_n}. \tag{9.12}$$

The distorted stator voltage introduces the corresponding harmonics in the stator current in the dq frame which can be described as follows:

$$\vec{i}_{s,dq} = \sum_{l=l_1}^{l_n} \vec{i}_{s,dq-l_n} = \vec{i}_{s,dq-l_1} + \vec{i}_{s,dq-l_2} + \cdots + \vec{i}_{s,dq-l_n}, \tag{9.13}$$

where l_1 is the fundamental component and l is the harmonic component.

The apparent power $S_{s,dq}$ can be represented under stator voltage distortion as:

$$\begin{aligned}
S_{s,dq} &= \sum_{l=l_1}^{l_n} \frac{3}{2}\left(\vec{v}_{s,dq-l_n} \cdot \vec{i}^{\,*}_{s,dq-l_n}\right) \\
&= \frac{3}{2}\left[\left(\vec{v}_{s,dq-l_1} \cdot \vec{i}^{\,*}_{s,dq-l_1}\right)\right] + \frac{3}{2}\left[\left(\vec{v}_{s,dq-l_2} \cdot \vec{i}^{\,*}_{s,dq-l_2}\right)\right] \\
&\quad + \cdots + \frac{3}{2}\left[\left(\vec{v}_{s,dq-l_n} \cdot \vec{i}^{\,*}_{s,dq-l_n}\right)\right].
\end{aligned} \tag{9.14}$$

In this way, the stator reactive $Q_{s,dq}$ and active $P_{s,dq}$ power are elements of the apparent power

$$\begin{aligned}
P_{s,dq} = P_s = \mathcal{R}e(S_{s,dq}) &= \mathcal{R}e\left[\sum_{l=l_1}^{l_n} \frac{3}{2}\left(\vec{v}_{s,dq-l_n} \cdot \vec{i}^{\,*}_{s,dq-l_n}\right)\right] \\
&= \mathcal{R}e\left\{\frac{3}{2}\left[\left(\vec{v}_{s,dq-l_1} \cdot \vec{i}^{\,*}_{s,dq-l_1}\right)\right]\right\}
\end{aligned}$$

$$+\mathcal{R}e\left\{\frac{3}{2}\left[\left(\vec{v}_{s,dq-l_2}\cdot\vec{i}^{*}_{s,dq-l_2}\right)\right]\right\}$$
$$+\cdots+\mathcal{R}e\left\{\frac{3}{2}\left[\left(\vec{v}_{s,dq-l_n}\cdot\vec{i}^{*}_{s,dq-l_n}\right)\right]\right\} \quad (9.15)$$
$$=\frac{3}{2}\left(v_{s,d-l_1}i_{s,d-l_1}+v_{s,q-l_1}i_{s,q-l_1}\right)$$
$$+\frac{3}{2}\left(v_{s,d-l_2}i_{s,d-l_2}+v_{s,q-l_2}i_{s,q-l_2}\right)$$
$$+\cdots+\frac{3}{2}\left(v_{s,d-l_n}i_{s,d-l_n}+v_{s,q-l_n}i_{s,q-l_n}\right),$$

$$\mathcal{Q}_{s,dq}=\mathcal{Q}_s=\mathcal{I}m(S_{s,dq})=\mathcal{I}m\left[\sum_{l=l_1}^{l_n}\frac{3}{2}\left(\vec{v}_{s,dq-l_n}\cdot\vec{i}^{*}_{s,dq-l_n}\right)\right]$$
$$=\mathcal{I}m\left\{\frac{3}{2}\left[\left(\vec{v}_{s,dq-l_1}\cdot\vec{i}^{*}_{s,dq-l_1}\right)\right]\right\}$$
$$+\mathcal{I}m\left\{\frac{3}{2}\left[\left(\vec{v}_{s,dq-l_2}\cdot\vec{i}^{*}_{s,dq-l_2}\right)\right]\right\}$$
$$+\cdots+\mathcal{I}m\left\{\frac{3}{2}\left[\left(\vec{v}_{s,dq-l_n}\cdot\vec{i}^{*}_{s,dq-l_n}\right)\right]\right\}$$
$$=\frac{3}{2}\left(v_{s,q-l_1}i_{s,d-l_1}-v_{s,d-l_1}i_{s,q-l_1}\right)$$
$$+\frac{3}{2}\left(v_{s,q-l_2}i_{s,d-l_2}-v_{s,d-l_2}i_{s,q-l_2}\right)$$
$$+\cdots+\frac{3}{2}\left(v_{s,q-l_n}i_{s,d-l_n}-v_{s,d-l_n}i_{s,q-l_n}\right), \quad (9.16)$$

where $\mathcal{I}m$ expresses the imaginary component and $\mathcal{R}e$ represents the real component of the complex number.

Hence, Eqs. (9.15) and (9.16) introduce the oscillations due to harmonic components in the elements of the apparent power of the stator.

9.2.2 Influence of distorted voltage in DC link voltage

In the DFIG operation, when the rotor side converter is linked to the DC side, it is assumed that the rotor voltage vector has only its fundamental components due to the control actions. Hence, $\vec{v}_{r,dq}=\vec{v}_{r,dq-l_1}$. The rotor current in the dq frame when the stator voltage is distorted can be expressed as:

$$\vec{i}_{r,dq}=\sum_{l=l_1}^{l_n}\vec{i}_{r,dq-l_n}=\vec{i}_{r,dq-l_1}+\vec{i}_{r,dq-l_2}+\cdots+\vec{i}_{r,dq-l_n}. \quad (9.17)$$

Thus, the active power of the rotor for any harmonic components can be represented as:

$$
\begin{aligned}
P_{r,dq} = \mathcal{R}e(S_{r,dq}) &= \mathcal{R}e\left[\sum_{l=l_1}^{l_n} \frac{3}{2}\left(\vec{v}_{r,dq-l_1} \cdot \vec{i}^{\,*}_{r,dq-l_n}\right)\right] \\
&= \mathcal{R}e\left\{\frac{3}{2}\left[\left(\vec{v}_{r,dq-l_1} \cdot \vec{i}^{\,*}_{r,dq-l_1}\right)\right]\right\} \\
&\quad + \mathcal{R}e\left\{\frac{3}{2}\left[\left(\vec{v}_{r,dq-l_1} \cdot \vec{i}^{\,*}_{r,dq-l_2}\right)\right]\right\} \\
&\quad + \cdots + \mathcal{R}e\left\{\frac{3}{2}\left[\left(\vec{v}_{r,dq-l_1} \cdot \vec{i}^{\,*}_{r,dq-l_n}\right)\right]\right\} \\
&= \frac{3}{2}\left(v_{r,d-l_1} i_{r,d-l_1} + v_{r,q-l_1} i_{r,q-l_1}\right) \\
&\quad + \frac{3}{2}\left(v_{r,d-l_1} i_{r,d-l_2} + v_{r,q-l_1} i_{r,q-l_2}\right) \\
&\quad + \cdots + \frac{3}{2}\left(v_{r,d-l_1} i_{r,d-l_n} + v_{r,q-l_1} i_{r,q-l_n}\right).
\end{aligned}
\tag{9.18}
$$

The rotor active power in Eq. (9.18) can be described as follows:

$$
\begin{aligned}
P_{r,dq} &= \overbrace{\frac{3}{2}\left(v_{r,d-l_1} i_{r,d-l_1} + v_{r,q-l_1} i_{r,q-l_1}\right)}^{P_{r,dq-l_1}} \\
&\quad \underbrace{+ \sum_{l=l_2}^{l_n} \frac{3}{2}\left(v_{r,d-l_1} i_{r,d-l_n} + v_{r,q-l_1} i_{r,q-l_n}\right)}_{P_{r,dq-l_n}},
\end{aligned}
\tag{9.19}
$$

where $P_{r,dq-l_1}$ and $P_{r,dq-l_n}$ are the active power of the rotor due to the fundamental component and all harmonic components greater than 2, respectively.

The power of the capacitor in the DC link (P_{cap}) can be represented as per the active power of the three-phase power CCG and the active power of the rotor when the power losses of the switches are disregarded. Hence, the power is given by

$$
\underbrace{P_{cap}}_{P_{cap-DC}+P_{cap-AC}} + \underbrace{P_{r,dq}}_{P_{r,dq-l_1}+P_{r,dq-l_n}} = \underbrace{P_{g,dq}}_{P_{g,dq-l_1}+P_{g,dq-l_n}},
\tag{9.20}
$$

where P_{cap-DC} and P_{cap-AC} are the DC and AC elements of the power of the capacitor in the DC link, respectively. The term P_{cap} in Eq. (9.20) can be rewritten considering the AC components:

$$
\begin{aligned}
P_{cap-AC} &= -P_{r,dq-l_n} + P_{g,dq-l_n} \\
&= -\sum_{l=l_2}^{l_n} \frac{3}{2}\left(v_{r,d-l_1} i_{r,d-l_n} + v_{r,q-l_1} i_{r,q-l_n}\right) \\
&\quad + \sum_{l=l_2}^{l_n} \frac{3}{2}\left(v_{g,d-l_n} i_{g,d-l_n} + v_{g,q-l_n} i_{g,q-l_n}\right).
\end{aligned}
\tag{9.21}
$$

Hence, Eq. (9.21) introduces the oscillations due to harmonic components in the active power of the rotor and grid, respectively.

9.3 DFIG rotor current control using MPRC-WM

This section introduces the DFIG rotor current control design by means of the MPRC in which the PWM is employed. So, in this chapter the technique is called MPRC-WM. For it, it is necessary to calculate the expanded model of the DFIG using the theory introduced in Chapter 8. The objective is to minimize the impact due to the distorted grid voltage in the stator current vector elements, where the matrices' and vectors' sizes are known as presented in Section 7.1 in its fundamental frequency and the perturbation in this model is the term $G_{gd,ir} d_{d,ir}(k)$. The mentioned perturbation is in accordance with the stator voltage distorted due to Eq. (3.13) ($\frac{v_s}{\omega_s} \cong \lambda_s$) or it can be seen in Eq. (9.8). In this way, each entry of the mentioned term $G_{gd,ir} d_{d,ir}(k)$ can be written as the inverse z-transform of $1/D(z)$. Hence, in steady state actions, $G_{gd,ir} d_{d,ir}(k)$, for every $k \geq k_0$, it can be described by means of the equation $D(q^{-1})(G_{gd,ir} d_{d,ir}(k)) = 0$, in which q^{-1} expresses the backward shift operator and k_0 represents the initial sampling instant.

The DFIG model in the dq frame in fundamental frequency using SVOC can be represented by using the rotor current vector as state and rotor voltage as input as in Eq. (7.2):

$$
\overbrace{\begin{bmatrix} i_{r,d}(k+1) \\ i_{r,q}(k+1) \end{bmatrix}}^{x_{ir}(k+1)} = \overbrace{\begin{bmatrix} 1 - \dfrac{R_r T}{\sigma L_r} & \omega_{sl} T \\ -\omega_{sl} T & 1 - \dfrac{R_r T}{\sigma L_r} \end{bmatrix}}^{A_{ad,ir}} \overbrace{\begin{bmatrix} i_{r,d}(k) \\ i_{r,q}(k) \end{bmatrix}}^{x_{ir}(k)}
+ \overbrace{\begin{bmatrix} \dfrac{T}{\sigma L_r} & 0 \\ 0 & \dfrac{T}{\sigma L_r} \end{bmatrix}}^{B_{bd,ir}} \overbrace{\begin{bmatrix} v_{r,d}(k) \\ v_{r,q}(k) \end{bmatrix}}^{u_{u,ir}(k)}
$$

$$+\overbrace{\begin{bmatrix} 0 & \dfrac{\omega_{sl} L_M T}{\sigma L_s L_r} \\ \dfrac{-\omega_{sl} L_m T}{\sigma L_s L_r} & 0 \end{bmatrix}}^{G_{gd,ir}} \overbrace{\begin{bmatrix} \lambda_s(k) \\ 0 \end{bmatrix}}^{d_{d,ir}(k)},$$

with $y_{ir}(k+1) = C_{cd,ir} x_{ir}(k+1)$ and $C_{cd,ir} = I$.

This model will be employed in the MPRC-WM design due to its simplicity when compared to Eq. (9.7). Additionally, the MPRC-WM technique has to be able to reject the disturbance due to the stator voltage distorted. In this context, the state space expression of the expanded model of the DFIG in the dq frame by means of the introduction of the polynomial can be presented in

$$\begin{aligned} x_{ee,ir}(k+1) &= A_{ee,ir} x_{ee,ir}(k) + B_{ee,ir} u_{xx,ir}(k), \\ y_{y,ir}(k) &= C_{ee,ir} x_{ee,ir}(k), \end{aligned} \tag{9.22}$$

where the indices ee, ir show a relationship to the expanded matrices, $x_{xx,ir}(k) = D(q^{-1}) x_{ir}(k)$, $u_{xx,ir}(k) = D(q^{-1}) u_{u,ir}(k)$,

$$\begin{aligned} y_{y,ir}(k+1) = &-d_1 y_{y,ir}(k) - d_2 y_{y,ir}(k-1) - \cdots - d_\rho y_{y,ir}(k-\rho+1) \\ &+ C_{cd,ir} A_{ad,ir} x_{xx,ir}(k) + C_{cd,ir} B_{bd,ir} u_{xx,ir}(k), \end{aligned} \tag{9.23}$$

as well as

$$x_{ee,ir}(k) = \begin{bmatrix} x_{xx,ir}(k)^T & y_{y,ir}(k)^T & \ldots & y_{y,ir}(k-\rho+1)^T \end{bmatrix}^T, \tag{9.24}$$

$$A_{ee,ir} = \begin{bmatrix} A_{ad,ir} & O \\ \hat{M}_{ir} & A_{pp,ir} \end{bmatrix} \hat{M}_{ir} = \begin{bmatrix} C_{cd,ir} A_{ad,ir} \\ O \end{bmatrix}, \tag{9.25}$$

$$A_{pp,ir} = \begin{bmatrix} -d_1 I & -d_2 I & \ldots & -d_{\rho-1} I & -d_\rho I \\ I & O & O & \ldots & O \\ 0 & I & O & \ldots & O \\ \vdots & \ddots & \ddots & \ddots & \vdots \\ O & O & \ldots & I & O \end{bmatrix}, \tag{9.26}$$

$$B_{ee,ir} = \begin{bmatrix} B_{bd,ir} \\ C_{cd,ir} B_{bd,ir} \\ O \\ \vdots \\ O \end{bmatrix} C_{ee,ir} = \begin{bmatrix} O & I & O & \ldots & O \end{bmatrix}, \tag{9.27}$$

where O and I represent zero and identity matrices, respectively.

The future rotor current vector components of the DFIG using its discrete representation of the expanded state model can be described as

$$\mathbf{I_{pr,ir}} = A_{pr,ir} x_{ee,ir}(k_i) + B_{pr,ir} \mathbf{V_{pr,ir}}, \tag{9.28}$$

where each term can be written as

$$\mathbf{I_{pr,ir}} = \left[x_{ee,ir}(k_i + 1|k_i)^T \quad \ldots \quad x_{ee,ir}(k_i + n_y|k_i)^T \right]^T, \tag{9.29}$$

$$\mathbf{V_{pr,ir}} = \left[u_{xx,ir}(k_i)^T \quad \ldots \quad u_{xx,ir}(k_i + n_u - 1)^T \right]^T, \tag{9.30}$$

$$A_{pr,ir} = \begin{bmatrix} A_{ee,ir} \\ A^2_{ee,ir} \\ \vdots \\ A^{n_p}_{ee,ir} \end{bmatrix}, \tag{9.31}$$

$$B_{pr,ir} = \begin{bmatrix} B_{ee,ir} & O & \ldots & O \\ A_{ee,ir} B_{ee,ir} & B_{ee,ir} & \ldots & O \\ A^2_{ee,ir} B_{ee,ir} & A_{ee,ir} B_{ee,ir} & \ldots & O \\ \vdots & \ddots & \ddots & \vdots \\ A^{n_y-1}_{ee,ir} B_{ee,ir} & A^{n_y-2}_{ee,ir} B_{ee,ir} & \ldots & A^{n_y-n_u}_{ee,ir} B_{ee,ir} \end{bmatrix}.$$

In Eq. (9.28), there is a new state space formulation for DFIG using rotor equations, in which the sampling instant is k_i and it always satisfies $k_i > 0$. The measurement of the state provides information about the system. Also $\mathbf{I_{pr,ir}}$ and $\mathbf{V_{pr,ir}}$ are the predicted states and control values for the state and input, respectively. It is assumed that $n_u \leq n_y$ and for $k \geq n_u$ all the control samples $u_{ee,ir}(k_i + k)$ become zero to reduce the computational cost.

The predicted control values $\mathbf{V_{pr,ir}}$ can be described as per the MPRC-WM using the cost function as presented in Eq. (9.32):

$$J_{pr,ir} = \mathbf{I_{pr,ir}}^T \overline{Q_{i,ir}} \mathbf{I_{pr,ir}} + \mathbf{V_{pr,ir}}^T \overline{R_{v,ir}} \mathbf{V_{pr,ir}}, \tag{9.32}$$

where $\overline{R_{v,ir}}$ describes a block diagonal matrix that has the $R_{v,ir} I$ form, in which $R_{v,ir}$ expresses a symmetric positive definite matrix and $\overline{Q_{i,ir}}$ expresses a block diagonal matrix that has the $Q_{i,ir} I$ form, in which $Q_{i,ir}$ represents a symmetric positive semidefinite matrix. Additionally, it is possible to add constraints as presented in Section 4.3.1, if necessary. So, the minimized cost function of Eq. (9.32) can be described as presented in Eq. (9.33):

$$\begin{aligned} J_{pr,ir} = {} & \mathbf{V_{pr,ir}}^T (B^T_{pr,ir} \overline{Q_{i,ir}} B_{pr,ir} + \overline{R_{v,ir}}) \mathbf{V_{pr,ir}} \\ & + 2 \mathbf{V_{pr,ir}}^T B^T_{pr,ir} \overline{Q_{i,ir}} A_{pr,ir} x_{ee,ir}(k_i) \\ & + x_{ee,ir}(k_i)^T A^T_{pr,ir} Q_{i,ir} A_{pr,ir} x_{ee,ir}(k_i). \end{aligned} \tag{9.33}$$

When there are no constraints, the optimal vector $\mathbf{V_{pr,ir}}$ can be represented by Eq. (9.33) as

$$\mathbf{V_{pr,ir}} = -(B_{pr,ir}^T \overline{Q_{i,ir}} B_{pr,ir} + \overline{R_{v,ir}})^{-1} B_{pr,ir}^T \overline{Q_{i,ir}} A_{pr,ir} x_{ee,ir}(k_i). \quad (9.34)$$

Extracting the values of $u_{ee,ir}$ from the vector $\mathbf{V_{pr,ir}}$ by means of $x_{xx,ir}(k) = D(q^{-1})x_{ir}(k)$ and $u_{xx,ir}(k) = D(q^{-1})u_{u,ir}(k)$, the input control signal to be inserted into the sample k_i is presented by:

$$u_{u,ir}(k_i) = u_{ee,ir}(k_i) - d_1 u_{u,ir}(k_i - 1) - \cdots - d_\rho u_{u,ir}(k_i - \rho). \quad (9.35)$$

When the MPRC-WM system operates by using a high sampling rate compared to the fundamental frequency of the DFIG, the past sample can be obtained with minimal error. In MPRC-WM, it will be employed to track a reference signal. Hence, the insertion of the reference signal into the expanded state space model is mandatory. In this context, the augmented output vector $x_{ee,ir}(k_i)$, after insertion of the reference signal, can be written as in Eq. (9.36) considering the error signal $e_{ir}(k)$. Thus, there is a new equation to represent $x_{ee,ir}(k_i)$, namely

$$x_{ee,ir}(k) = \begin{bmatrix} x_{xx,ir}(k)^T & e_{ir}(k)^T & \ldots & e_{ir}(k - \rho + 1)^T \end{bmatrix}^T. \quad (9.36)$$

The error signal can be represented in Eq. (9.37). And, it is the difference of the output signal $(y_{ir}(k_i))$ and the reference signal $(r_{ir}(k_i))$,

$$e_{ir}(k) = y_{ir}(k_i) - r_{ir}(k_i). \quad (9.37)$$

In this book, the rotor current of the DFIG is controlled by means of the voltage vectors that are computed at each sampling time by the MPRC-WM under distorted stator voltage. When the voltage vector is computed by means of MPRC-WM, Eq. (9.35) allows accomplishing this task. Hence, the rotor voltage vector $(\vec{v}_{r,\alpha\beta r})$ in the $\alpha\beta r$ frame can be calculated by using $\vec{v}_{r,\alpha\beta r} = \vec{v}_{r,dq} e^{j(\theta_s - \theta_r)}$. Vector $\vec{v}_{r,\alpha\beta r}$ can be the input of the PWM technique of the three-phase power converter connected to the rotor. So, the rotor current vector elements will reach the reference vector and the stator current will have no influence due to the stator voltage distortions.

9.3.1 Criterion for choosing polynomial $D(z)$

The criterion for the polynomial $D_{ir}(z)$ is in accordance with the type of the reference signal, and more details can be seen in Section 8.2.1.1. In the rotor current control for DFIG by using MPRC-WM, the objective is to reach a constant reference due to the representation of the DFIG in the dq frame. Hence, just the term $(1 - z^{-1})$, which represents a low-pass filter, can be used. In this way, the polynomial is inserted into the representation of the DFIG to create the expanded model that can be seen in

$$D_{ir}(z) = (1 - z^{-1}). \tag{9.38}$$

The rotor current control of DFIG using MPRC-WM can be expressed by the scheme presented in Fig. 9.1, and its basic code can be seen in Algorithm 9.1.

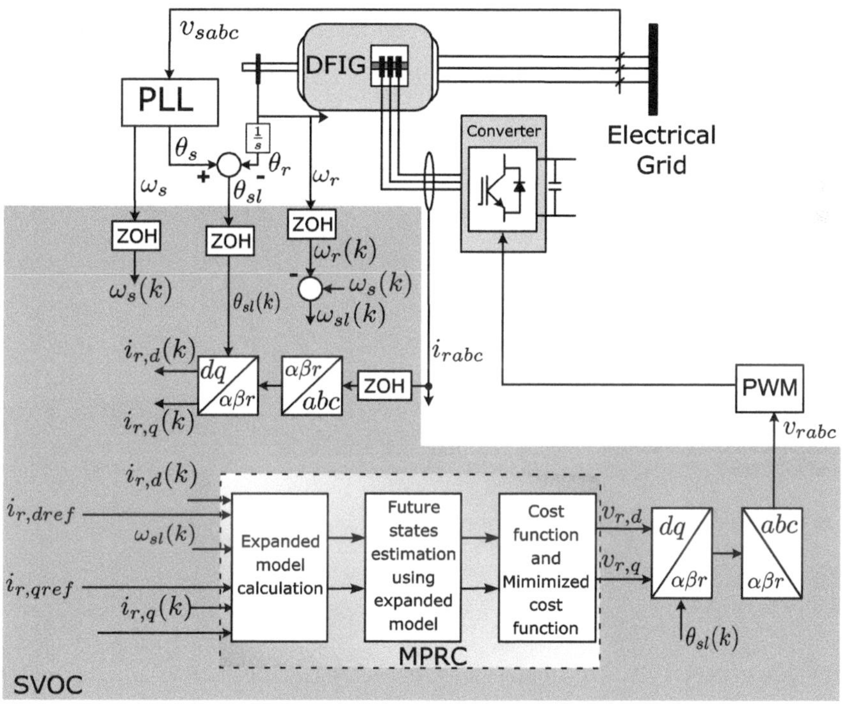

FIGURE 9.1 Diagram of rotor current control of DFIG using MPRC-WM.

Algorithm 9.1 DFIG rotor current control using MPRC-WM algorithm

---

while Controller Active **do**
  $output, e(k) \leftarrow$ System measurements
  $[x_{xx,ir}(k),\quad e] = \text{func}_1(x_{ir}(k), \vec{e}(k), D(z), x_{xx,ir})$
  Compute: $[A_{ad,ir},\quad B_{bd,ir},\quad C_{cd,ir}]$
  $[A_{ee,ir},\quad B_{ee,ir}] = \text{func}_2(A_{ad,ir}, B_{bd,ir}, C_{cd,ir}, D(z))$
  $[A_{pr,ir},\quad B_{pr,ir}] = \text{func}_3(A_{ee,ir}, B_{ee,ir}, n_u, n_u)$
  $\overline{Q_{i,ir}}, \overline{R_{v,ir}} \leftarrow$ Assign values
  $x_{ee,ir}(k) = \text{func}_4(x_{xx,ir}(k))$
  $\mathbf{V_{pr,ir}} = (x_{ee,ir}(k), A_{pr,ir}, B_{pr,ir}, \overline{Q_{i,ir}}, \overline{R_{v,ir}})$
  $input(k) = \text{func}_5(\mathbf{V_{pr,ir}}, D(z), input)$
end while

9.3.2 Experimental results

This section introduces the experimental results obtained in the setup using the data presented in Table A.1 and the rotor speed is maintained by the DC motor at 1700 rpm. The horizons are $n_u = 2$, $n_y = 3$, $\overline{Q_{i,ir}} = 2e^{12} \cdot I$, and $\overline{R_{v,ir}} = 5e^6 \cdot I$.

9.3.2.1 *Normal operation tests*

The behavior of the rotor current vector elements in the dq frame during several step tests is depicted in Fig. 9.2. It can be noticed that the MPRC-WM calculates the rotor voltage vector elements necessary for the rotor current vector components to reach their references. The details of the rotor current response, in which the settling time is about 2 ms, can be seen in Fig. 9.3, and in this test $i_{r,d}$ remains constant. The oscillations occur due the switching of the inverter.

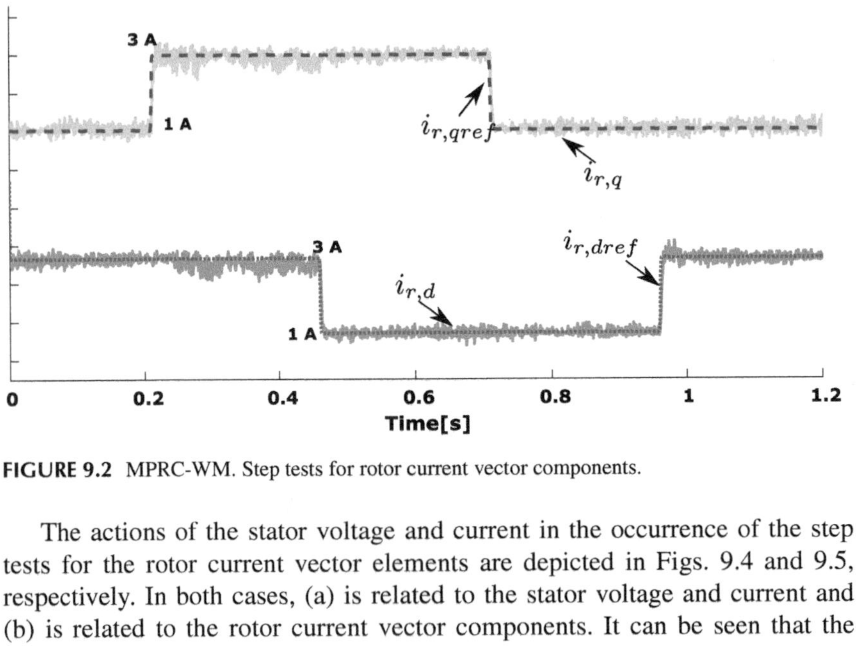

FIGURE 9.2 MPRC-WM. Step tests for rotor current vector components.

The actions of the stator voltage and current in the occurrence of the step tests for the rotor current vector elements are depicted in Figs. 9.4 and 9.5, respectively. In both cases, (a) is related to the stator voltage and current and (b) is related to the rotor current vector components. It can be seen that the stator current increases its amplitude when $i_{r,q}$ changes its value, or the stator current changes its angle when $i_{r,d}$ changes its value due to the fact the rotor current vector in the dq frame is related to the stator reactive and active power in accordance with the model representation.

In the last test, when the DFIG operates by changing the values of the speed and the rotor current vector elements are constant at $i_{r,q} = i_{r,d} = 1$ A, the actions of the speed of the rotor, the rotor current in the $\alpha\beta r$ frame, and the rotor current vector components in the dq frame are depicted in Figs. 9.6(a)–(c). It can be noticed in Fig. 9.6 that the rotor current control by using MPRC-WM reaches the references.

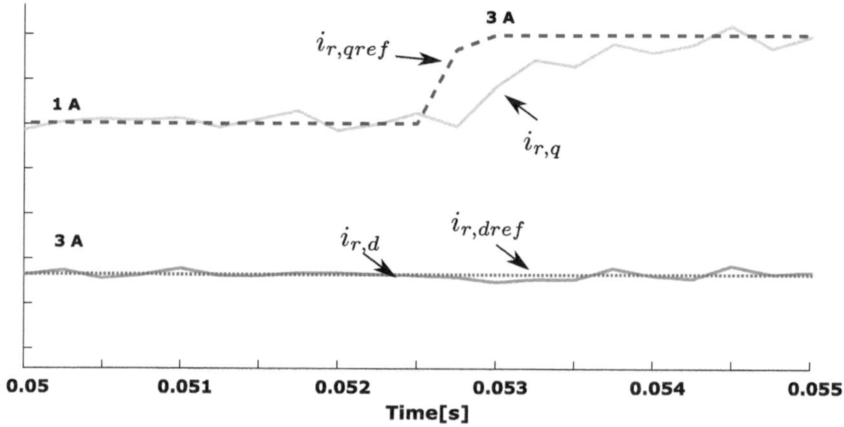

FIGURE 9.3 MPRC-WM. Details of the step test for $i_{r,q}$.

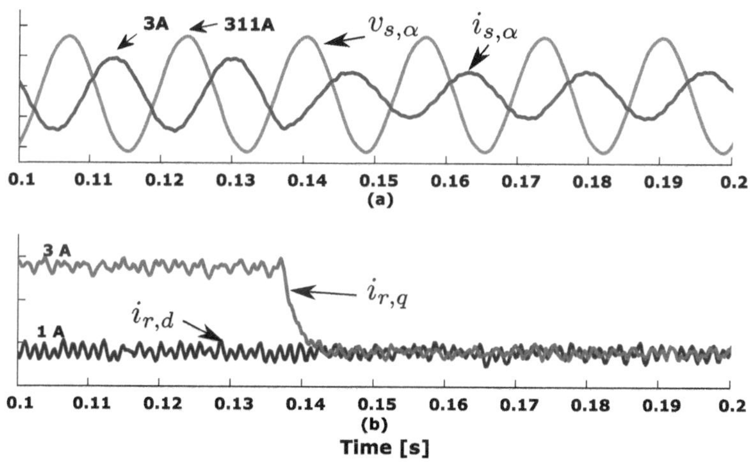

FIGURE 9.4 MPRC-WM. Detail of the step test for $i_{r,q}$: (a) stator voltage and current, (b) rotor current vector components.

9.3.2.2 Tests under stator distorted voltage operation

This section introduces the experimental results obtained under distorted voltage operation. In this case, the fundamental frequency is 60 Hz plus 7% of the 5th harmonic. The distorted voltage peak value is 245 V. In this test the reference of the rotor current vector is maintained constant, and we analyze the capability of the MPRC-WM to reject the harmonic component in the stator current.

The rotor current vector components, stator voltage and current, and the THD of stator current (3.97%) during the rotor current control using MPRC-WM operation are depicted in Figs. 9.7(a), 9.7(b), and 9.8, respectively. The MPRC-WM reaches the references of the rotor current vector, and the stator

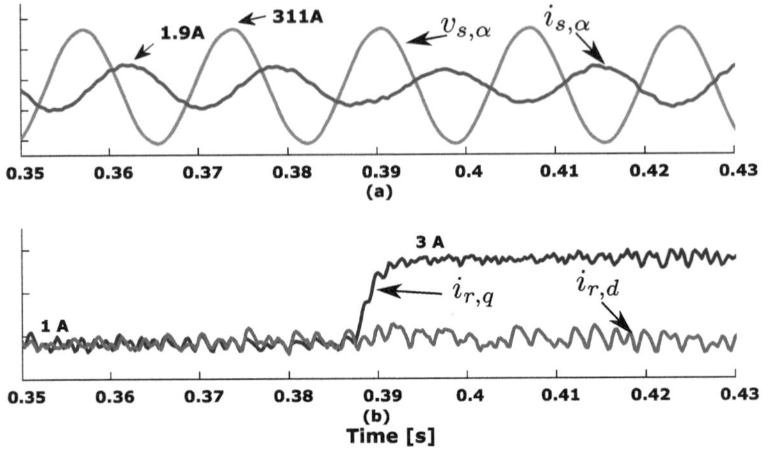

FIGURE 9.5 MPRC-WM. Details of the step test for $i_{r,d}$: (a) stator voltage and current, (b) rotor current vector components.

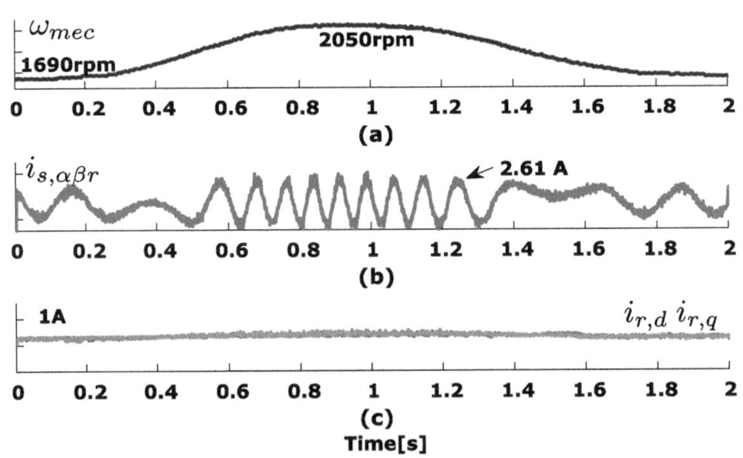

FIGURE 9.6 MPRC-WM. Test using several-speed operation: (a) speed of the rotor, (b) rotor current in rotor frame, and (c) rotor current vector components in the dq frame.

current has minor harmonic components during the occurrence of nonsinusoidal voltage. Hence, THD is nearly 3 times smaller when compared to the THD of the stator current (11.36%) using MPC-WM (Figs. 9.9 and 9.10). However, the rotor current vector components have more oscillations when compared to the MPC-WM due to the fact the MPRC-WM effort is to reject the disturbance when the stator voltage has harmonic components. In this way, the main objective is to minimize the mentioned distortion in the stator current. So, this comparison attests the better performance of the rotor current control using MPRC-WM against the conventional MPC-WM.

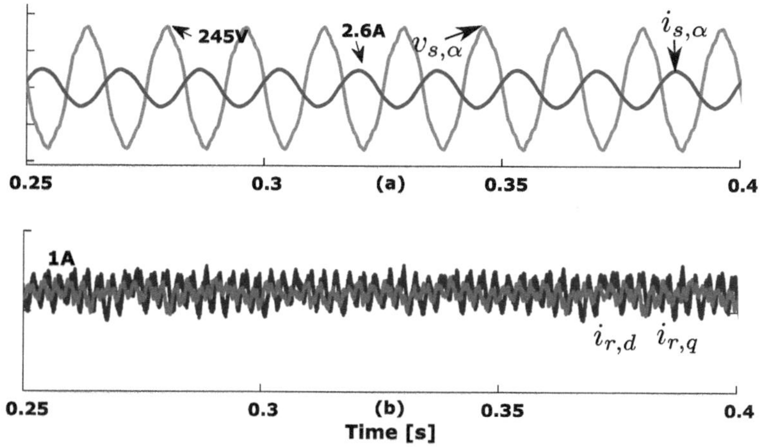

FIGURE 9.7 MPRC-WM. Test under voltage distorted by 7% of the 5th harmonic component: (a) stator voltage and current, (b) rotor current vector components.

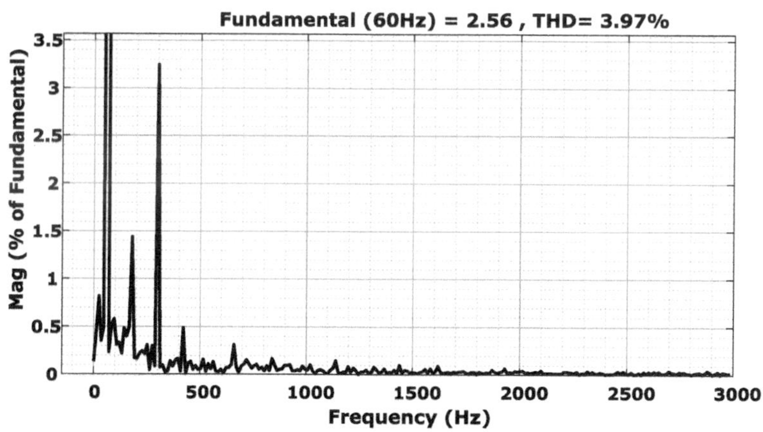

FIGURE 9.8 MPRC-WM. THD and harmonics components of stator current.

9.4 Three-phase power CCG model under grid distorted voltage

The three-phase power CCG model can be represented by Eq. (2.70) ($\vec{v}_{inv,dq} = R_g \vec{i}_{g,dq} + L_g \frac{d\vec{i}_{g,dq}}{dt} + \vec{v}_{g,dq} + j\omega_g L_g \vec{i}_{g,dq}$), in which its components can be expressed by means of Eqs. (2.71) and (2.72). Hence, using the Laplace trans-

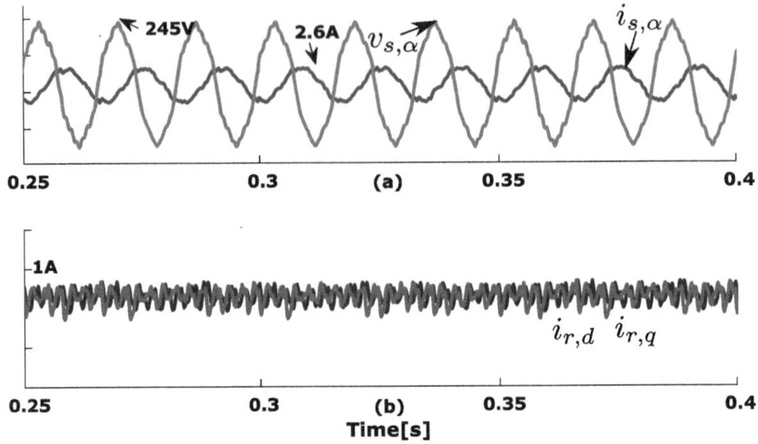

FIGURE 9.9 MPC-WM. Test under voltage distorted by 7% of the 5th harmonic component: (a) stator voltage and current, (b) rotor current vector components.

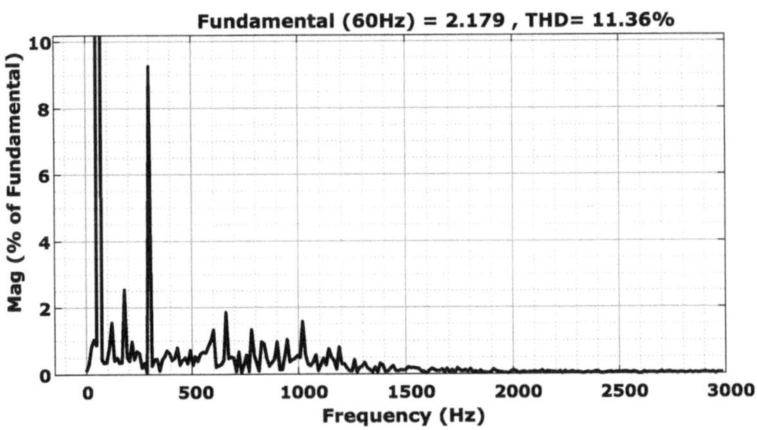

FIGURE 9.10 MPC-WM. THD and harmonics components of the stator current.

formation, Eq. (2.70) can be expressed as follows:

$$
\begin{aligned}
&\vec{v}_{inv,dq}(s) = R_g \vec{i}_{g,dq}(s) + L_g s \vec{i}_{g,dq}(s) + \vec{v}_{g,dq}(s) + j\omega_g L_g \vec{i}_{g,dq}(s),\\
&\frac{\vec{i}_{g,dq}(s)}{\vec{v}_{inv,dq}(s) - \vec{v}_{g,dq}(s)} = \frac{1}{R_g + L_g s + j\omega_g L_g}.
\end{aligned}
\tag{9.39}
$$

Eq. (9.39) introduces the open-loop model, and it can be noticed that the grid voltage distortions impact the model of the CCG.

9.4.1 Influence of distorted voltage in the active and reactive power representation

When the grid voltage in the dq frame is under the presence of the harmonic components, in the converter voltage $\vec{v}_{inv,dq}$ we can neglect the harmonic components, so it can be only represented by the fundamental component due to the control action. Hence, we can write it as

$$\vec{v}_{inv,dq} = \vec{v}_{inv,dq-l_1}. \tag{9.40}$$

The distorted grid voltage can be expressed as

$$\vec{v}_{g,dq} = \sum_{l=l_1}^{l_n} \vec{v}_{g,dq-l_n} = \vec{v}_{g,dq-l_1} + \vec{v}_{g,dq-l_2} + \cdots + \vec{v}_{g,dq-l_n}. \tag{9.41}$$

The distorted grid voltage introduces the corresponding harmonics in the grid current in the dq frame, which can be described as

$$\vec{i}_{g,dq} = \sum_{l=l_1}^{l_n} \vec{i}_{g,dq-l_n} = \vec{i}_{g,dq-l_1} + \vec{i}_{g,dq-l_2} + \cdots + \vec{i}_{g,dq-l_n}, \tag{9.42}$$

where l_1 is the fundamental frequency and l denotes the harmonics components.

The apparent power $S_{g,dq}$ can be represented under grid voltage distortion as follows:

$$\begin{aligned} S_{g,dq} &= \sum_{l=l_1}^{l_n} \frac{3}{2}\left(\vec{v}_{g,dq-l_1} \cdot \vec{i}^{\,*}_{g,dq-l_n}\right) \\ &= \frac{3}{2}\left[\left(\vec{v}_{g,dq-l_1} \cdot \vec{i}^{\,*}_{g,dq-l_1}\right)\right] + \frac{3}{2}\left[\left(\vec{v}_{g,dq-l_2} \cdot \vec{i}^{\,*}_{g,dq-l_2}\right)\right] \\ &\quad + \cdots + \frac{3}{2}\left[\left(\vec{v}_{g,dq-l_n} \cdot \vec{i}^{\,*}_{g,dq-l_n}\right)\right]. \end{aligned} \tag{9.43}$$

The grid active $P_{g,dq}$ and reactive $Q_{g,dq}$ power are components of the apparent power as follows:

$$\begin{aligned} P_{g,dq} &= \mathcal{R}e(S_{g,dq}) \\ &= \mathcal{R}e\left[\sum_{l=l_1}^{l_n} \frac{3}{2}\left(\vec{v}_{g,dq-l_n} \cdot \vec{i}^{\,*}_{g,dq-l_n}\right)\right] \\ &= \mathcal{R}e\left\{\frac{3}{2}\left[\left(\vec{v}_{g,dq-l_1} \cdot \vec{i}^{\,*}_{g,dq-l_1}\right)\right]\right\} \\ &\quad + \mathcal{R}e\left\{\frac{3}{2}\left[\left(\vec{v}_{g,dq-l_2} \cdot \vec{i}^{\,*}_{g,dq-l_2}\right)\right]\right\} \end{aligned} \tag{9.44}$$

$$
\begin{aligned}
&+\cdots+\mathcal{R}e\left\{\frac{3}{2}\left[\left(\vec{v}_{g,dq-l_n}\cdot\vec{i}^{\,*}_{g,dq-l_n}\right)\right]\right\}\\
&=\frac{3}{2}\left(v_{g,d-l_1}i_{g,d-l_1}+v_{g,q-l_1}i_{g,q-l_1}\right)\\
&+\frac{3}{2}\left(v_{g,d-l_2}i_{g,d-l_2}+v_{g,q-l_2}i_{g,q-l_2}\right)\\
&+\cdots+\frac{3}{2}\left(v_{g,d-l_n}i_{g,d-l_n}+v_{g,q-l_n}i_{g,q-l_n}\right),
\end{aligned}
$$

$$
\begin{aligned}
Q_{g,dq}&=\mathcal{I}m(S_{g,dq})\\
&=\mathcal{I}m\left[\sum_{l=l_1}^{l_n}\frac{3}{2}\left(\vec{v}_{g,dq-l_n}\cdot\vec{i}^{\,*}_{g,dq-l_n}\right)\right]\\
&=\mathcal{I}m\left\{\frac{3}{2}\left[\left(\vec{v}_{g,dq-l_1}\cdot\vec{i}^{\,*}_{g,dq-l_1}\right)\right]\right\}\\
&+\mathcal{I}m\left\{\frac{3}{2}\left[\left(\vec{v}_{g,dq-l_2}\cdot\vec{i}^{\,*}_{g,dq-l_2}\right)\right]\right\}\\
&+\cdots+\mathcal{I}m\left\{\frac{3}{2}\left[\left(\vec{v}_{g,dq-l_n}\cdot\vec{i}^{\,*}_{g,dq-l_n}\right)\right]\right\}\\
&=\frac{3}{2}\left(v_{g,q-l_1}i_{g,d-l_1}-v_{g,d-l_1}i_{g,q-l_1}\right)\\
&+\frac{3}{2}\left(v_{g,q-l_2}i_{g,d-l_2}-v_{g,d-l_2}i_{g,q-l_2}\right)\\
&+\cdots+\frac{3}{2}\left(v_{g,q-l_n}i_{g,d-l_n}-v_{g,d-l_n}i_{g,q-l_n}\right),
\end{aligned}
\tag{9.45}
$$

where $\mathcal{I}m$ represents the imaginary component and $\mathcal{R}e$ expresses the real component of the complex number. The active power represented by Eq. (9.44) is the power injected to the DC link using the grid side converter when the power losses of the switches are neglected.

Eqs. (9.43)–(9.45) demonstrate the influence of the distorted voltage in the active and reactive power.

9.5 Three-phase power CCG current control using MPRC-WM

This section introduces the three-phase CCG current control design by means of the MPRC-WM. The objective is to minimize the impact to the distorted grid voltage in the grid current. For this, it is necessary to calculate the expanded model of the three-phase power CCG, where the matrices' and vectors' sizes are known as presented in Section 7.3. In this application, three-phase power CCG model in the $\alpha\beta$ frame considering the disturbance can be represented by using the grid current vector as state and converter minus grid voltage vectors as input (Eq. (7.28)):

$$\overbrace{\begin{bmatrix} i_{g,\alpha}(k+1) \\ i_{g,\beta}(k+1) \end{bmatrix}}^{x_g(k+1)} = \overbrace{\begin{bmatrix} 1-\dfrac{R_gT}{L_g} & 0 \\ 0 & 1-\dfrac{R_gT}{L_g} \end{bmatrix}}^{A_{ad,g}} \overbrace{\begin{bmatrix} i_{g,\alpha}(k) \\ i_{g,\beta}(k) \end{bmatrix}}^{x_g(k)} + \\ + \overbrace{\begin{bmatrix} \dfrac{T}{L_g} & 0 \\ 0 & \dfrac{T}{L_g} \end{bmatrix}}^{B_{bd,g}} \overbrace{\begin{bmatrix} v_{inv,\alpha}(k) - v_{g,\alpha-l_1}(k) \\ v_{inv,\beta}(k) - v_{g,\beta-l_1}(k) \end{bmatrix}}^{u_{u,g}(k)} + \mu(k),$$

where $y_g(k+1) = C_{cd,g}x_g(k+1)$ and the harmonic components of the grid voltage are represented by the disturbance $\mu(k) = \sum_{l=l_2}^{l_n} \vec{v}_{g,\alpha\beta-l_n}$, $C_{cd,g}$ is the identity matrix I, $v_{g,\alpha-l_1}(k)$ and $v_{g,\beta-l_1}(k)$ are the elements of the grid voltage vector in the $\alpha\beta$ frame in fundamental frequency.

In this context, the state space representation of the expanded model of the three-phase CCG in the $\alpha\beta$ frame after the introduction of a polynomial can be presented in Eq. (9.46). In steady state, the disturbance ($\mu(k)$), for every $k \geq k_0$, can be expressed using the equation $D(q^{-1})(\mu(k)) = 0$, in which q^{-1} represents the backward shift operator and k_0 is the initial sampling instant:

$$\begin{aligned} x_{ee,g}(k+1) &= A_{ee,g}x_{ee,g}(k) + B_{ee,g}u_{xx,g}(k), \\ y_{y,g}(k) &= C_{ee,g}x_{ee,g}(k), \end{aligned} \tag{9.46}$$

where the indices ee, g relate to the expanded matrices, $x_{xx,g}(k){=}D(q^{-1})x_g(k)$, $u_{xx,g}(k) = D(q^{-1})u_{u,g}(k)$, and

$$\begin{aligned} y_{y,g}(k+1) = &-d_1y_{y,g}(k) - d_2y_{y,g}(k-1) - \cdots - d_\rho y_{y,g}(k-\rho+1) \\ &+ C_{cd,g}A_{ad,g}x_{xx,g}(k) + C_{cd,g}B_{bd,g}u_{xx,g}(k), \end{aligned} \tag{9.47}$$

as well as

$$x_{ee,g}(k) = \begin{bmatrix} x_{xx,g}(k)^T & y_{y,g}(k)^T & \dots & y_{y,g}(k-\rho+1)^T \end{bmatrix}^T, \tag{9.48}$$

$$A_{ee,g} = \begin{bmatrix} A_{ad,g} & O \\ \hat{M}_g & A_{pp,g} \end{bmatrix} \hat{M}_g = \begin{bmatrix} C_{cd,g}A_{ad,g} \\ O \end{bmatrix}, \tag{9.49}$$

$$A_{pp,g} = \begin{bmatrix} -d_1 I & -d_2 I & \dots & -d_{\rho-1} I & -d_\rho I \\ I & O & O & \dots & O \\ 0 & I & O & \dots & O \\ \vdots & \ddots & \ddots & \ddots & \vdots \\ O & O & \dots & I & O \end{bmatrix}, \tag{9.50}$$

$$B_{ee,g} = \begin{bmatrix} B_{bd,g} \\ C_{cd,g} B_{bd,g} \\ O \\ \vdots \\ O \end{bmatrix} C_{ee,g} = \begin{bmatrix} O & I & O & \dots & O \end{bmatrix}, \tag{9.51}$$

where O and I represent zero and identity matrices, respectively.

The future grid current vector components by using its discrete representation of the augmented state model can be described as follows:

$$\mathbf{I_{pr,g}} = A_{pr,g} x_{ee,g}(k_i) + B_{pr,g} \mathbf{V_{pr,g}}, \tag{9.52}$$

where each term can be represented as:

$$\mathbf{I_{pr,g}} = \left[x_{ee,g}(k_i + 1|k_i)^T \quad \dots \quad x_{ee,g}(k_i + n_y|k_i)^T \right]^T, \tag{9.53}$$

$$\mathbf{V_{pr,g}} = \left[u_{xx,g}(k_i)^T \quad \dots \quad u_{xx,g}(k_i + n_u - 1)^T \right]^T, \tag{9.54}$$

$$A_{pr,g} = \begin{bmatrix} A_{ee,g} \\ A_{ee,g}^2 \\ \vdots \\ A_{ee,g}^{n_p} \end{bmatrix}, \tag{9.55}$$

$$B_{pr,g} = \begin{bmatrix} B_{ee,g} & O & \dots & O \\ A_{ee,g} B_{ee,g} & B_{ee,g} & \dots & O \\ A_{ee,g}^2 B_{ee,g} & A_{ee,g} B_{ee,g} & \dots & O \\ \vdots & \ddots & \ddots & \vdots \\ A_{ee,g}^{n_y-1} B_{ee,g} & A_{ee,g}^{n_y-2} B_{ee,g} & \dots & A_{ee,g}^{n_y-n_u} B_{ee,g} \end{bmatrix}.$$

In Eq. (9.52), there is a new state space formulation for three-phase CCG, in which the sampling instant is k_i and it always satisfies $k_i > 0$. The measurement of the state provides information about the system. The $\mathbf{I_{pr,g}}$ and $\mathbf{V_{pr,g}}$ are the predicted states and control values for the state and input, respectively. It is assumed that $n_u \leq n_y$, and for $k \geq n_u$ all samples $u_{ee,g}(k_i + k)$ become zero to reduce the computational cost. Additionally, as the electrical grid is under normal operation conditions, the assumption for low values of n_y, $\vec{v}_{inv,\alpha\beta}(k) \cong \vec{v}_{inv,\alpha\beta}(k+1) \cong \cdots \cong \vec{v}_{inv,\alpha\beta}(k+n_y-1)$ due to the sampling time T is much smaller than the period of the grid.

The predicted control values $\mathbf{V_{pr,g}}$ can be calculated by MPRC-WM using the cost function as presented below:

$$J_{pr,g} = \mathbf{I_{pr,g}}^T \overline{Q_{i,g}} \mathbf{I_{pr,g}} + \mathbf{V_{pr,g}}^T \overline{R_{v,g}} \mathbf{V_{pr,g}}, \tag{9.56}$$

where $\overline{R_{v,g}}$ represents a block diagonal matrix that has the $R_{v,g}I$ form, in which $R_{v,g}$ expresses a symmetric positive definite matrix and $\overline{Q_{i,g}}$ describes a block diagonal matrix that has the $Q_{i,g}I$ form, in which $Q_{i,g}$ describes a symmetric positive semidefinite matrix. Additionally, it is possible to add constraints, as presented in Section 4.3.1. So, the minimized cost function in Eq. (9.56) can be rewritten as

$$\begin{aligned} J_{pr,g} = {} & \mathbf{V_{pr,g}}^T (B_{pr,g}^T \overline{Q_{i,g}} B_{pr,g} + \overline{R_{v,g}}) \mathbf{V_{pr,g}} \\ & + 2\mathbf{V_{pr,g}}^T B_{pr,g}^T \overline{Q_{i,g}} A_{pr,g} x_{ee,g}(k_i) \\ & + x_{ee,g}(k_i)^T A_{pr,g}^T Q_{i,g} A_{pr,g} x_{ee,g}(k_i). \end{aligned} \tag{9.57}$$

When there are no constraints, the optimal vector $\mathbf{V_{pr,g}}$ can be represented by means of Eq. (9.57) as given by

$$\mathbf{V_{pr,g}} = -(B_{pr,g}^T \overline{Q_{i,g}} B_{pr,g} + \overline{R_{v,g}})^{-1} B_{pr,g}^T \overline{Q_{i,g}} A_{pr,g} x_{ee,g}(k_i). \tag{9.58}$$

Using the extracted values of $u_{ee,g}$ from the vector $\mathbf{V_{pr,g}}$, as well as $x_{xx,g}(k) = D(q^{-1}) x_g(k)$ and $u_{xx,g}(k) = D(q^{-1}) u_{u,g}(k)$, the input control signal to be inserted at the instant k_i can be described as follows:

$$u_{u,g}(k_i) = u_{ee,g}(k_i) - d_1 u_{u,g}(k_i - 1) - \cdots - d_\rho u_{u,g}(k_i - \rho). \tag{9.59}$$

When the MPRC-WM system operates by using a high sampling rate compared to the fundamental frequency of the CCG, the past sample can be estimated with minimal error. In the MPRC-WM, it will be employed to track a reference signal. Hence, the insertion of the reference signal into the expanded state space representation of the CCG is mandatory. In this context, the augmented output vector $x_{ee,g}(k_i)$, after the insertion of the reference signal, can be written as in Eq. (9.60) considering the error signal $e_g(k)$. Hence, there is a new equation to represent $x_{ee,g}(k_i)$ as follows:

$$x_{ee,g}(k) = \left[x_{xx,g}(k)^T \quad e_g(k)^T \quad \ldots \quad e_g(k-\rho+1)^T \right]^T. \tag{9.60}$$

The error signal can be represented in Eq. (9.61). And, it is the difference of the output signal $(y_g(k_i))$ and the reference signal $(r_g(k_i))$, namely

$$e_g(k) = y_g(k_i) - r_g(k_i). \tag{9.61}$$

The references of the grid current vector elements can be reached by using the grid reactive and active power references and the grid voltage vector elements in the fundamental frequency using Eq. (7.37) as follows:

$$\begin{bmatrix} i_{g,\alpha ref} \\ i_{g,\beta ref} \end{bmatrix} = \frac{2}{3[v_{g,\alpha-l_1}^2(k) + v_{g,\beta-l_1}^2(k)]} \begin{bmatrix} v_{g,\alpha-l_1}(k) & v_{g,\beta-l_1}(k) \\ v_{g,\beta-l_1}(k) & -v_{g,\alpha-l_1}(k) \end{bmatrix} \begin{bmatrix} P_{gref} \\ Q_{gref} \end{bmatrix}.$$

In this book, the grid current of the three-phase power CCG is controlled by means of the voltage vectors that are computed at each sampling time by the MPRC-WM. When the voltage vector is computed as per the MPRC-WM, Eq. (9.59) allows accomplishing this task. Hence, the voltage vector of the CCG components will be the input of the PWM algorithm of the three-phase power converter. The grid current control using MPRC-WM can be represented by means of the scheme presented in Fig. 9.11, and its basic code can be seen in Algorithm 9.2. The fundamental frequency component and the angle of the grid voltage are necessary to calculate the grid current vector reference using

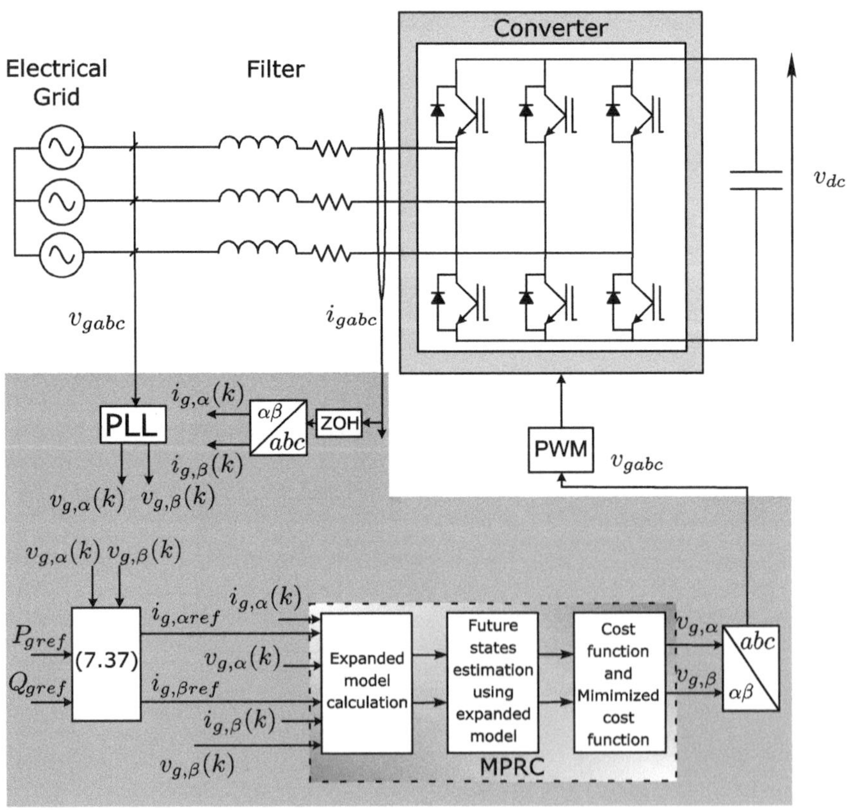

FIGURE 9.11 Diagram of grid current control using MPRC-WM.

Eq. (7.37). Hence, the PLL is able to operate under distorted voltage due harmonics components, and one can reach this objective. So, the grid current vector will reach the reference vector with no influence due to the grid voltage distortions because of harmonics components.

Algorithm 9.2 CCG grid current control using MPRC-WM algorithm

while Controller Active **do**
- $output, e(k) \leftarrow$ System measurements
- $[x_{xx,g}(k), \quad e] = \text{func}_1(x_g(k), \vec{e}(k), D(z), x_{xx,g})$
- Compute: $[A_{ad,g}, \quad B_{bd,g}, \quad C_{cd,g}]$
- $[A_{ee,g}, \quad B_{ee,g}] = \text{func}_2(A_{ad,g}, B_{bd,g}, C_{cd,g}, D(z))$
- $[A_{pr,g}, \quad B_{pr,g}] = \text{func}_3(A_{ee,g}, B_{ee,g}, n_u, n_u)$
- $\overline{Q_{i,g}}, \overline{R_{v,g}} \leftarrow$ Assign values
- $x_{ee,g}(k) = \text{func}_4(x_{xx,g}(k))$
- $\mathbf{V_{pr,g}} = (x_{ee,g}(k), A_{pr,g}, B_{pr,g}, \overline{Q_{i,g}}, \overline{R_{v,g}})$
- $input(k) = \text{func}_5(\mathbf{V_{pr,g}}, D_g(z), input)$

end while

9.5.1 Criterion for choosing the polynomial $D_g(z)$

The criterion for the polynomial $D_g(z)$ is in accordance with the type of the reference signal; more details can be seen in Section 8.2.1.1. In the grid current control for the three-phase CCG by using MPRC-WM, the objective is to reach a sinusoidal reference signal in the fundamental frequency of the electrical grid due to the model in the $\alpha\beta$ frame. In this way, the polynomial is inserted into the representation of the CCG to create the expanded model that can be written as

$$D_g(z) = (1 - z^{-1}). \tag{9.62}$$

9.5.2 Experimental results

This section presents the experimental results obtained in the considered setup by using the data presented in Table B.1. The horizons are $n_u = 2$, $n_y = 3$, $\overline{Q_{i,g}} = 3 \cdot 10^6 I$, and $\overline{R_{v,g}} = 5 \cdot 10^1 1I$.

9.5.2.1 Grid under normal operation

The behavior of step tests for grid power control as the grid active power (P_g) varies from 500 to 750 W and the grid reactive power (Q_g) varies from 0 to −300 var is depicted in Figs. 9.12(a) and 9.12(b). The settling time is about 1.5 ms, as can be seen in Fig. 9.13(a), and in this test Q_g remains constant at 0 var as depicted in Fig. 9.13(b).

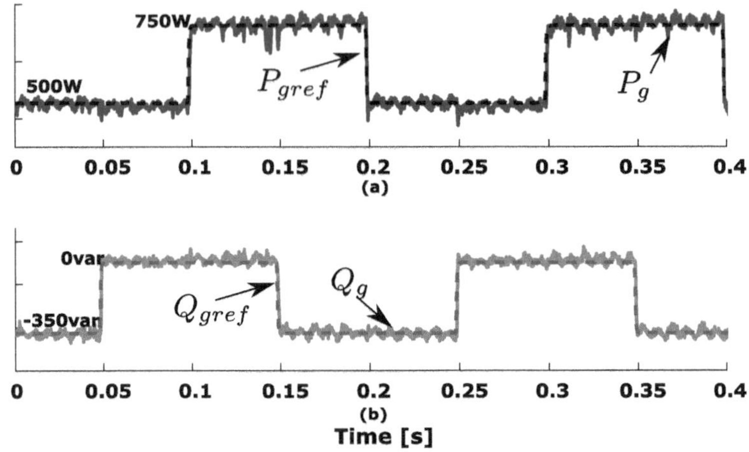

FIGURE 9.12 MPRC-WM. Step tests for active and reactive power: (a) active power, (b) reactive power.

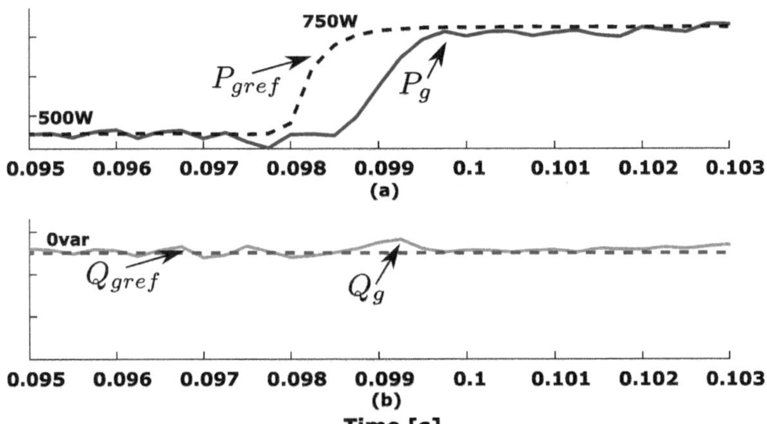

FIGURE 9.13 MPRC-WM. Details of the step test for active power: (a) active power, (b) reactive power.

In this way, it can be noticed that the MPRC-WM calculates the grid voltage vector necessary for the grid current vector components and the power to reach their references. The actions of the grid current vector elements are depicted in Figs. 9.14(a) and 9.14(b). The grid current reference vector is calculated by means of the power as presented in Eq. (7.37). The oscillations occur due to the switching of the inverter.

The grid current and voltage during the reactive and active grid power step tests are depicted in Figs. 9.15 and 9.16, respectively. It can be noticed that the grid voltage remains constant, as it is an infinite bus, and the grid current increases its amplitude when the active power increases its value, and the grid

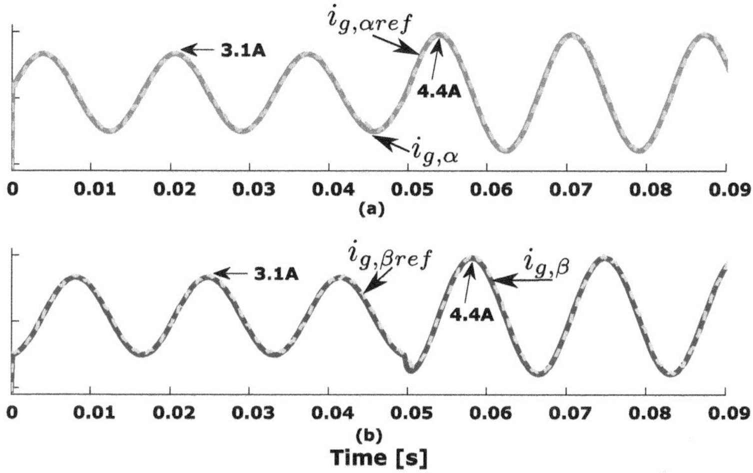

FIGURE 9.14 MPRC-WM. Step tests for grid current vector components: (a) real component of grid current vector, (b) imaginary component of grid current vector.

current changes its angle when the reactive power changes its values in accordance with the model representation.

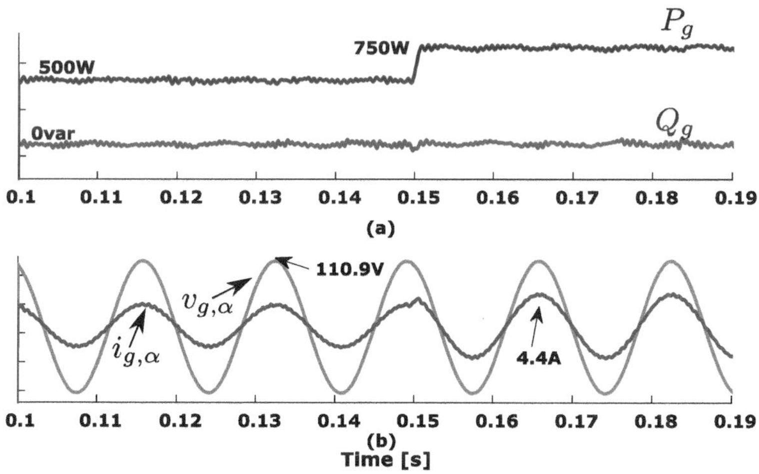

FIGURE 9.15 MPRC-WM. Grid voltage and current action during step test of active power: (a) reactive and active power, (b) grid voltage and current.

9.5.2.2 Grid under distorted voltage operation

This section introduces the experimental results obtained under distorted voltage operation. In this case, the harmonics components in the grid under distorted voltage operation are presented in Table 9.1. The grid distorted voltage peak value is 111 V. In this case, the reference of the grid current vector is maintained

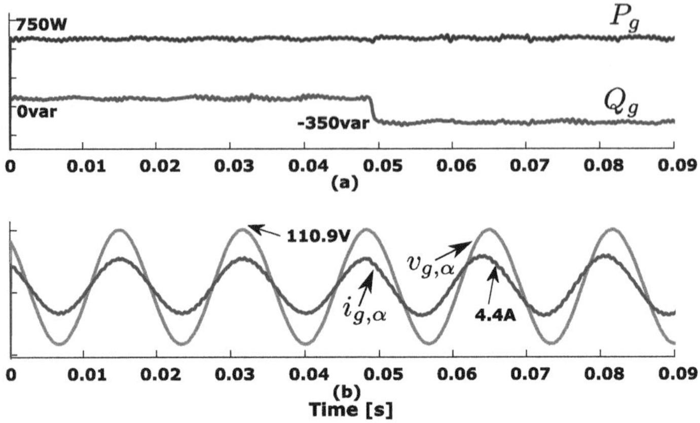

FIGURE 9.16 MPRC-WM. Grid voltage and current action during a step test of reactive power: (a) reactive and active power, (b) grid voltage and current.

constant, and we analyze the capability of the MPRC-WM to reject the harmonic components in the grid current. It is important to mention that the PLL that operates under distorted voltage was employed in both controllers to calculate the grid voltage vector components in their fundamental frequency.

TABLE 9.1 Distorted voltage harmonics components.

Harmonic component	Magnitude
1st – 60 Hz	100.00%
5th	3.94%
7th	3.15%
11th	2.36%
13th	1.50%
17th	1.10%
19th	0.70%

The grid voltage and current action during the step test of active power from 500 to 750 W, while the reactive power is maintained at 0 var during the distorted grid voltage, is presented in Figs. 9.17(a) and 9.17(b), respectively. The MPRC-WM reaches the references of the power using grid current vector references. The grid current has minor harmonic components during the voltage distortions using MPRC-WM (see Fig. 9.18, where THD is 3.95%) when compared to operating using MPC-WM (Section 7.3.2), where the same mentioned test is depicted in Fig. 9.19, and THD is 4.85% as depicted in Fig. 9.20. Hence, the THD using MPRC-WM is nearly 20% smaller when compared to THD using the MPC-WM. However, the reactive and active power has oscillations due to the fact the controller effort rejects the disturbance when the stator voltage has a harmonic component. So, this comparison shows the better performance of the grid current control using MPRC-WM when compared to the conventional MPC-WM.

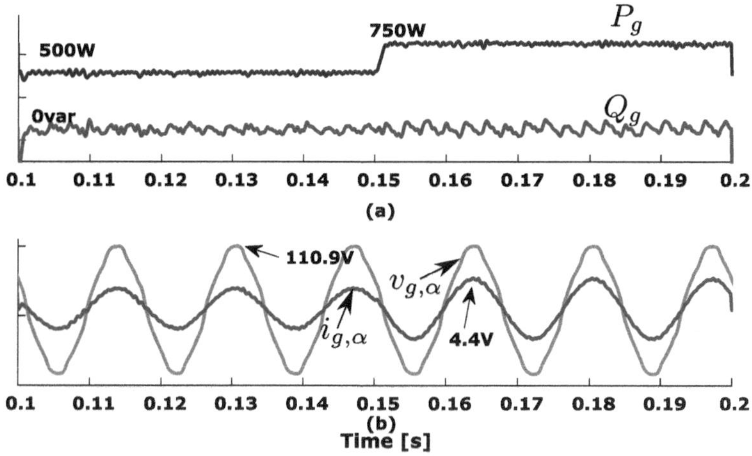

FIGURE 9.17 MPRC-WM. Grid voltage and current action during a step test of active power during the distorted grid voltage: (a) reactive and active power, (b) grid voltage and current.

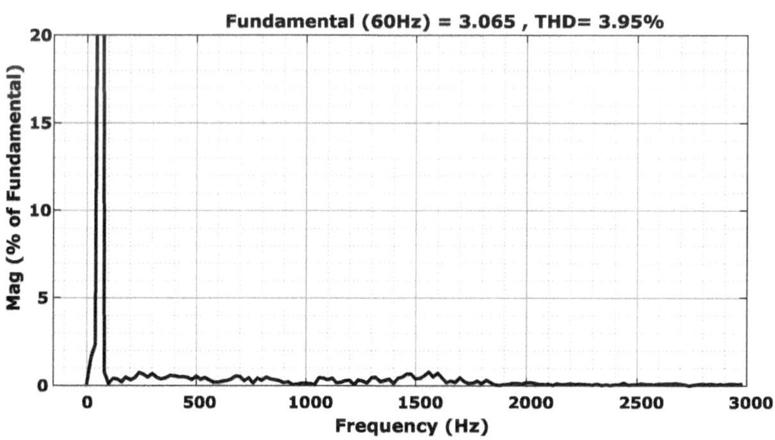

FIGURE 9.18 MPRC-WM. THD of stator current during the distorted grid voltage.

9.6 Summary

This chapter describes the rotor current control for DFIG and the grid current control for CCG using the MPRC-WM technique in normal operation and under distorted voltage conditions. The MPRC-WM employs the discrete representation of the mentioned systems, and the repetitive control actions are inserted

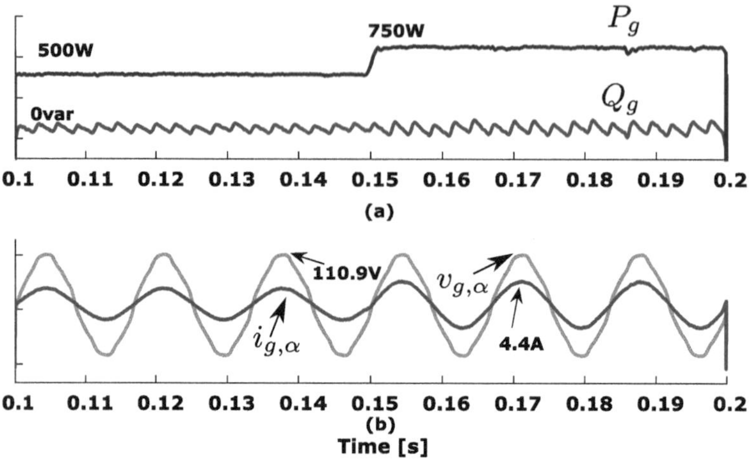

FIGURE 9.19 MPC-WM. Grid voltage and current action during a step test of active power during the distorted grid voltage: (a) reactive and active power, (b) grid voltage and current.

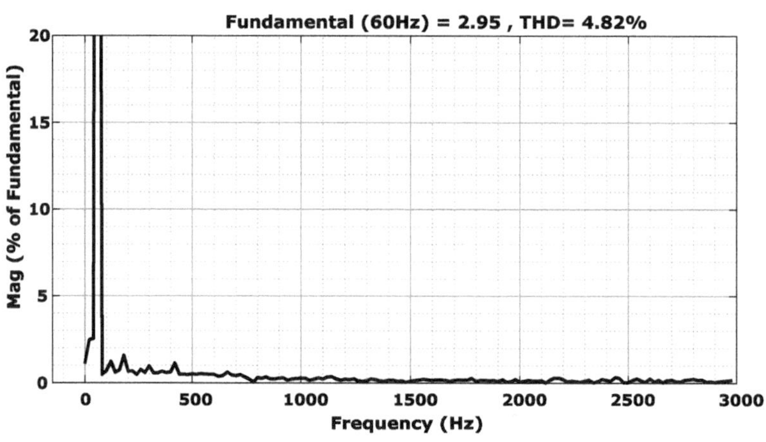

FIGURE 9.20 MPC-WM. THD of stator current during the distorted grid voltage.

into this technique. In this way, the DFIG is represented in the dq frame and the CCG is expressed in the $\alpha\beta$ frame. In both cases, the MPRC-WM permitted to reach the references using constant switching frequency due to the PWM and in accordance with the presented experimental results. Additionally, this technique has superior performance when compared to the MPC-WM and when the grid voltage is distorted by harmonics elements. In this case, the objective is to mitigate the distortions in the current of the stator of DFIG or current of the CCG. Hence, this type of scheme can be interesting for DFIG or CCG operating under grid voltage distortions.

Chapter 10

Finite position set phase-locked loop operating under nonideal grid voltages

Alfeu J. Sguarezi Filho[a], Fernando Lino[a], and Rogério V. Jacomini[b]
[a]*Federal University of ABC, Santo André, SP, Brazil,* [b]*Federal Institute of São Paulo, Hortolândia, SP, Brazil*

10.1 PLL fundamentals

The phase-locked loop (PLL) aims to calculate an output signal in which its phase is synchronized with the phase of an entry signal. The structure of the PLL is composed by three blocks and each one has a specific function [121]. They can be described as follows:

- The phase detector (PD) enables an output signal that is proportional to the variation between the entry signal v and the signal produced by the PLL v' phase.
- The loop filter (LF) has a low pass filter or PI and it can reduce the high frequency AC elements from the output of the PD.
- The voltage-controlled oscillator (VCO) enables an output AC signal in which its frequency is dislocated from the central frequency ω_{ff}. The VCO is related to the input voltage of the LF.

The basic structure of the PLL block diagram is depicted in Fig. 10.1.

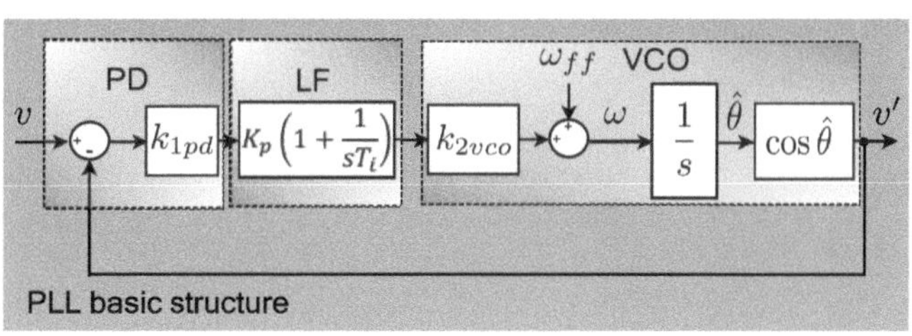

FIGURE 10.1 Basic structure of the PLL block diagram.

The closed loop transfer function was developed in [43] using the small signal model and the following conditions were assumed:

- For small values of $\Delta\theta$, the element $\sin(\Delta\theta) \cong \Delta\theta$.

https://doi.org/10.1016/B978-0-32-390964-8.00019-1

- The values $k_{1pd} = k_{2vco} = 1$.
- The $E(s) = V(s) - V'(s) \cong \theta(s) - \hat{\theta}(s)$, where $E(s)$, $V(s)$ and $V'(s)$ are the Laplace transforms of the error $e = v - v' \cong \theta - \hat{\theta}$, $v = \sin(\theta)$, and $v' = \cos(\hat{\theta})$, respectively.
- The term $V_{vco}(s) = \left(k_p + \frac{k_i}{s}\right) E(s)$, $k_i = \frac{k_p}{T_i}$.
- The term $\frac{1}{s} V_{vco}(s) = \hat{\theta}(s)$.

Hence, the closed-loop transfer function of the PLL considering the mentioned assumptions can be described as follows:

$$H_{cl}(s) = \frac{\hat{\theta}(s)}{\theta(s)} = \frac{k_i + k_p s}{k_i + k_p s + s^2}. \tag{10.1}$$

The normalized second-order transfer function can be expressed as

$$H_{cl}(s) = \frac{\hat{\theta}(s)}{\theta(s)} = \frac{\omega_n \xi 2s + \omega_n^2}{\omega_n^2 + \omega_n \xi 2s + s^2}, \tag{10.2}$$

where the gains of the PI using [43,122] can be computed as

$$\begin{aligned} \xi &= \frac{\sqrt{k_p^2/k_i}}{2}, \\ \omega_n &= \sqrt{k_i}. \end{aligned} \tag{10.3}$$

An alternative way to design the gains of the PLL using symmetrical optimum method was developed in [123].

10.1.1 PLL for three-phase systems

The PLL for three-phase systems in the dq frame (SRF-PLL) can be achieved using the scheme depicted in Fig. 10.1. The same assumptions as in the last section are being employed. This type of PLL is very important to power generation systems [124], and Fig. 10.2 depicts their block diagram.

In this way, the measured voltages of the balanced grid in a three-phase system are transformed to the dq frame using Eqs. (2.3) and (2.7) as follows:

$$\begin{bmatrix} v_{g,\alpha} \\ v_{g,\beta} \end{bmatrix} = \begin{bmatrix} 1 & -\frac{1}{2} & -\frac{1}{2} \\ 0 & \frac{\sqrt{3}}{2} & -\frac{\sqrt{3}}{2} \end{bmatrix} \begin{bmatrix} v_{g,a} \\ v_{g,b} \\ v_{g,c} \end{bmatrix} \tag{10.4}$$

and

$$\begin{bmatrix} v_{g,d} \\ v_{g,q} \end{bmatrix} = \begin{bmatrix} \cos(\theta_g) & \sin(\theta_g) \\ -\sin(\theta_g) & \cos(\theta_g) \end{bmatrix} \begin{bmatrix} v_{g,\alpha} \\ v_{g,\beta} \end{bmatrix}. \tag{10.5}$$

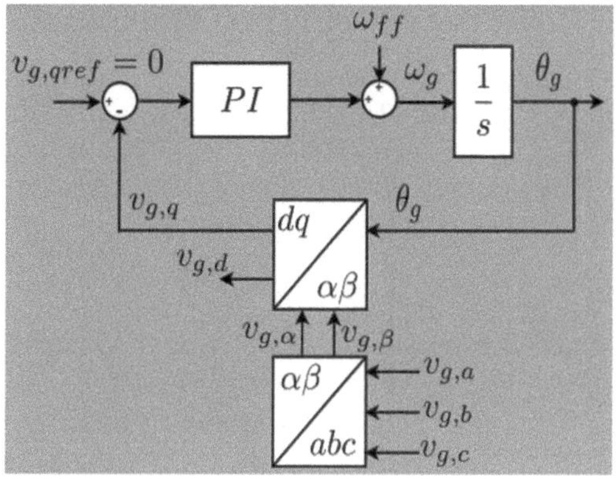

FIGURE 10.2 PLL for three-phase systems' block diagram.

When the PLL matches queue frequency ω_g, the components of the grid voltage vector $v_{g,d}$ and $v_{g,q}$ will appear as constant signals. This goal is achieved by adjusting the PI action based on the quadrature component $v_{g,q}$ until it is equal to zero. In this way, the PI changes the angular position of the output of the PLL in order to align the voltage vector with the direct axis. The frequency ω_{ff} allows increasing the stabilization of the PLL specially at start-up.

10.2 Representation of grid voltage disturbances

The disturbances in electrical grid occur due to the transient grid faults and due to the presence of nonlinear loads. In this book the electrical grid voltages are considered as being composed of its fundamental sequence component plus a harmonic component, which can be negative ($l < 0$) or positive ($l > 0$). The positive sequence in the fundamental frequency of the voltage vector interrelates with either a negative or positive sequence nth-order component. In this way, the three-phase voltages considering null the phase angle, for simplicity, can be described as

$$v_{g,abc} = \begin{bmatrix} v_{g,a} \\ v_{g,b} \\ v_{g,c} \end{bmatrix} = V^{+1} \begin{bmatrix} \cos(\omega_g t) \\ \cos(\omega_g t - \frac{2\pi}{3}) \\ \cos(\omega_g t + \frac{2\pi}{3}) \end{bmatrix} + V^l \begin{bmatrix} \cos(l\omega_g t) \\ \cos(l\omega_g t - \frac{2\pi}{3}) \\ \cos(l\omega_g t + \frac{2\pi}{3}) \end{bmatrix}, \quad (10.6)$$

where V^{+1} is the peak value of the fundamental frequency element of the positive sequence and V^l is the peak value of the lth-order harmonic element. The zero sequence elements are disregarded as it is assumed that the generator operates using three-wired connection. Additionally, the synchronization of any current with the zero-sequence voltage component is not employed in this book.

Eq. (10.6) in the $\alpha\beta$ frame using Eq. (2.2) can be represented as

$$\vec{v}_{g,\alpha\beta} = [T_{abc}]v_{g,abc} = V^{+1}\begin{bmatrix}\cos(\omega_g t)\\ \sin(\omega_g t)\end{bmatrix} + V^{l}\begin{bmatrix}\cos(l\omega_g t)\\ \sin(l\omega_g t)\end{bmatrix}, \tag{10.7}$$

where $[T_{abc}] = \begin{bmatrix}1 & -\frac{1}{2} & -\frac{1}{2}\\ 0 & \frac{\sqrt{3}}{2} & -\frac{\sqrt{3}}{2}\end{bmatrix}$.

Eq. (10.6) in the dq frame using Eqs. (2.7) and (10.7) can be represented as

$$\vec{v}_{g,dq} = [T_{dq}]\vec{v}_{g,\alpha\beta} = V^{+1}\begin{bmatrix}\cos(\omega_g t-\theta)\\ \sin(\omega_g t-\theta)\end{bmatrix} + V^{l}\begin{bmatrix}\cos(l\omega_g t-\theta)\\ \sin(l\omega_g t-\theta)\end{bmatrix}, \tag{10.8}$$

where $[T_{dq}] = \begin{bmatrix}\cos(\theta_g) & \sin(\theta_g)\\ -\sin(\theta_g) & \cos(\theta_g)\end{bmatrix}$ and θ_g is the position of the dq frame computed by means of the grid voltage vector elements.

It can be noticed that the second term of Eq. (10.8) represents a disturbance in the calculated elements of the grid voltage vector $v_{d,g}$ and $v_{q,g}$. In this way, the SRF-PLL tends to match the frequency ω_g of the positive sequence components of the grid voltage vector $v_{g,d}^{+}$ and $v_{g,q}^{+}$ subjected to the harmonic components,

$$\vec{v}_{g,dq} = [T_{dq}]\vec{v}_{g,\alpha\beta} = V^{+1}\begin{bmatrix}1\\ 0\end{bmatrix} + V^{l}\begin{bmatrix}\cos((l-1)\omega_g t)\\ \sin((l-1)\omega_g t)\end{bmatrix}. \tag{10.9}$$

In this way, the components of the grid voltage vector will no longer behave like DC signals due to the presence of harmonic terms when Eq. (10.9) is analyzed. Hence, when the amplitude of the harmonic components increases, the oscillation presented by the angle θ_g also increases, contributing to an increase of the error in the phase estimation performed by the SRF-PLL.

An asymmetrical grid fault can be represented using Eq. (10.9) when $l = -1$. Hence, it can be noticed that the oscillating term with double the fundamental frequency is associated with this type of operation and it can be represented as

$$\vec{v}_{g,dq} = [T_{dq}]\vec{v}_{g,\alpha\beta} = V^{+1}\begin{bmatrix}1\\ 0\end{bmatrix} + V^{-1}\begin{bmatrix}\cos(-2\omega_g t)\\ \sin(-2\omega_g t)\end{bmatrix}. \tag{10.10}$$

In this way, the oscillation presented by the angle θ_g increases, contributing to an increase of the error in the phase estimation performed by the SRF-PLL due to its bandwidth not being effective to reject this type of disturbance. Figs. 10.3(a)–(d) depict the experimental result of the performance of the SRF-PLL during an unbalanced voltage sag. In this way, one can noticed the ripple due the $2\omega_g$ element during the sag in the grid voltages elements in the dq frame

(Fig. 10.3(b)), grid frequency (Fig. 10.3(c)), and in the grid angle (Fig. 10.3(d)). Hence, this type of disturbance degrades the performance of the SRF-PLL.

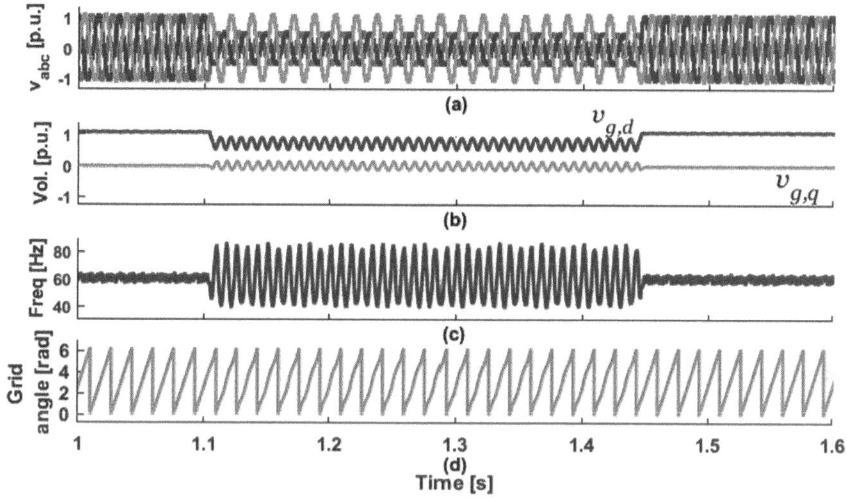

FIGURE 10.3 Experimental result of SRF-PLL test for unbalanced voltage sag: (a) three-phase voltage, (b) grid voltages elements in the dq frame, (c) grid frequency, and (d) grid angle.

10.3 Finite position set PLL operation under grid disturbances

This section introduces the finite position set PLL (FPS-PLL) together with dual second order generalized integrator (DSOGI) and moving average filter (MAF) operation during voltage sags or when the grid voltage is distorted by harmonic elements. In this way, the DSOGI permits computing the positive sequence of the grid voltage and the MAF mitigates the harmonic components of the input of FPS-PLL. In this book, the cited solution is called DSOGI-MAF-FPS-PLL.

10.3.1 Representation of the DSOGI

The DSOGI was developed in [125]. It employs adaptive filters based on the second-order generalized integrator to prospect the instantaneous symmetrical components method. In this way, the DSOGI implementation can be reached using quadrature signal generator (QSG) as per the representation below [43]:

$$\begin{aligned} \vec{v}^{+}_{g,\alpha\beta} &= [T^{+}_{\alpha\beta}]\vec{v}_{g,\alpha\beta} \Rightarrow [T^{+}_{\alpha\beta}] = \frac{1}{2}\begin{bmatrix} 1 & -e^{-j\pi/2} \\ e^{-j\pi/2} & 1 \end{bmatrix}, \\ \vec{v}^{-}_{g,\alpha\beta} &= [T^{-}_{\alpha\beta}]\vec{v}_{g,\alpha\beta} \Rightarrow [T^{-}_{\alpha\beta}] = \frac{1}{2}\begin{bmatrix} 1 & e^{-j\pi/2} \\ -e^{-j\pi/2} & 1 \end{bmatrix}. \end{aligned} \tag{10.11}$$

The term $e^{-j\pi/2}$ can be implemented employing the second-order adaptive filters [43] based on an SOGI (SOGI-QSG). This method allows acquiring from any sinusoidal input signal two quadrature output signals and it has the capacity to attenuate to the output the distortions due to the harmonic components of the input.

Fig. 10.4 depicts the structure of the DSOGI and it can be noticed that two SOGI-QSGs are employed. Hence, they are responsible for generating the quadrature and direct axis elements using the input vector.

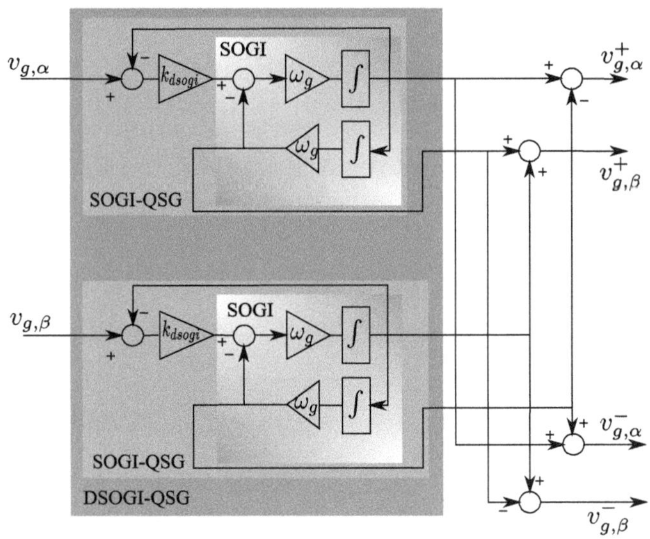

FIGURE 10.4 Block diagram of structure of the DSOGI.

The transfer function of the positive-sequence element discovered by the DSOGI using the unbalance input voltage vector and Eq. (10.11) can be represented as:

$$\begin{aligned}\vec{v}^{+}_{g,\alpha\beta} &= \begin{bmatrix} D_{dsogi}(s) & -Q_{dsogi}(s) \\ Q_{dsogi}(s) & D_{dsogi}(s) \end{bmatrix} \vec{v}_{g,\alpha\beta}, \\ \vec{v}^{+}_{g,\alpha\beta} &= \frac{1}{2}\frac{k_{dsogi}\omega_g}{s^2 + k_{dsogi}s\omega_g + \omega_g^2}\begin{bmatrix} s & -\omega_g \\ \omega_g & s \end{bmatrix} \vec{v}_{g,\alpha\beta},\end{aligned} \tag{10.12}$$

where k_{dsogi} is a constant, $D_{dsogi}(s) = \dfrac{k_{dsogi}s\omega_g}{s^2 + k_{dsogi}s\omega_g + \omega_g^2}$, and $Q_{dsogi}(s) = \dfrac{k^2_{dsogi}\omega_g}{s^2 + k_{dsogi}s\omega_g + \omega_g^2}$. The calculation of the transfer functions $D_{dsogi}(s)$ and $Q_{dsogi}(s)$ can be seen in detail in [43].

10.3.2 Representation of the MAF

The MAF is a filtering technique based on finite impulse response (FIR) filters and it can perform as an ideal low pass filter in determined terms. The PLL based on MAF was developed in [126]. Although this filter is easy to implement, it can have a high computational cost.

The representation of the output $\bar{x}_{maf}$ using the MAF in the continuous time domain can be achieved using the input expression x_{maf} as follows:

$$\bar{x}_{maf} = \frac{1}{T_{maf}} \int_{t-T_{maf}}^{t} x(\tau_{maf}) d\tau_{maf}. \tag{10.13}$$

The T_{maf} expresses the length of the window, and the MAF transfer function using Eq. (10.13) can be described as follows:

$$H_{maf}(s) = \frac{\bar{x}_{maf}(s)}{x(s)} = \frac{1 - e^{-T_{maf}s}}{T_{maf}s}. \tag{10.14}$$

To achieve the steady state actions, the MAF expressed using Eq. (10.14) needs the same time of its window length. The response of the MAF decreases when the window length increases.

The magnitude and phase of Eq. (10.14) can be achieved substituting $s = j\omega$ and considering some manipulations. Hence, it can be represented as follows:

$$\begin{aligned} H_{maf}(j\omega) &= \frac{1 - e^{-T_{maf}j\omega}}{T_{maf}j\omega} = \frac{1 - \cos(T_{maf}\omega) + j\sin(T_{maf}\omega)}{T_{maf}j\omega}, \\ |H_{maf}(j\omega)| &= \left| \frac{\sin(T_{maf}\omega/2)}{(T_{maf}\omega)/2} \right|, \\ H_{maf}(j\omega)\angle &= (-T_{maf}\omega)/2. \end{aligned} \tag{10.15}$$

It can be seen in (10.15) that the MAF has null gain at frequencies $f = n_{maf}/T_{maf}$ ($n_{maf} = 1, 2, 3, 4, \ldots$) and unity gain at null frequency in hertz. Hence, the MAF retains the frequency elements of integer multiples of $(1/T_{maf})$ in hertz, and one passes the DC element.

The MAF employed in practice requires the discrete time representation. In this way, in this book it is assumed that the window length of the filter includes N_{maf} samples (N_{maf} is an integer) of the input signal in which $T_{maf} = N_{maf}T$ and T is the sampling time. In this way, the discrete equation of MAF can be represented as

$$\bar{x}(k)_{maf} = \frac{1}{N_{maf}} \sum_{i_{maf}=0}^{N_{maf}-1} \bar{x}_{maf}(k - i_{maf}). \tag{10.16}$$

Eq. (10.16) in the z domain can be expressed as

$$\bar{X}(z)_{maf} = \left(\frac{1}{N_{maf}} \sum_{i_{maf}=0}^{N_{maf}-1} z^{i_{maf}} \right) X_{maf}(z),$$
$$\bar{X}(z)_{maf} = \frac{1}{N_{maf}} \frac{1 - z^{-N_{maf}}}{1 - z^{-1}} X_{maf}(z). \quad (10.17)$$

10.3.3 Finite position set PLL

The FPS-PLL was developed in [127,128] for permanent magnet synchronous machines and using the one-cycle Fourier FPS-PLL in [129], in which these methods are based on MPC-FCS and the cited solutions replace the PI controller in the SRF-PLL that can be employed to the grid synchronization. Hence, the minimized cost function was employed in this chapter to reach the optimal grid angle position from a finite number of grid angle positions that vary between 0 and 2π.

The FPS-PLL processes the elements of the positive sequence of the grid voltage vector in the dq frame. Hence, the equation of the elements can be expressed using Eq. (2.7) as:

$$v_{g,d}^{+} = v_{g,\alpha}^{+} \cos\theta_g + v_{g,\beta}^{+} \sin\theta_g,$$
$$v_{g,q}^{+} = v_{g,\beta}^{+} \cos\theta_g - v_{g,\alpha}^{+} \sin\theta_g, \quad (10.18)$$

where θ_g represents the angle of the grid positive sequence voltage.

The grid voltage-orientation is employed in the FPS-PLL and it is reached when the direct axis of the dq frame is aligned with the grid voltage vector, as presented in Section 3.2.2. Hence, $v_{s,d} = v_g = \left|\vec{v}_{g,dq}\right|$ and $v_{g,q} = 0$.

The FPS-PLL predicts the behavior of the grid voltage components using the representation of Eq. (10.18) in discrete manner using the Euler's method and it is represented as:

$$v_{g,q}^{+}(k, i_{fps}, j_{fps}) = v_{g,\beta}^{+}(k) \cos\theta_g(k, i_{fps}, j_{fps}) - v_{g,\alpha}^{+}(k) \sin\theta_g(k, i_{fps}, j_{fps}),$$
$$v_{g,d}^{+}(k, i_{fps}, j_{fps}) = v_{g,\alpha}^{+}(k) \cos\theta_g(k, i_{fps}, j_{fps}) + v_{g,\beta}^{+}(k) \sin\theta_g(k, i_{fps}, j_{fps}). \quad (10.19)$$

The angle $\theta_g(k, i_{fps}, j_{fps})$ is obtained using the initial position $\theta_{in}(k, i_{fps})$ that permits getting a finite number of angular positions, using two nested iteration loops with indexes i_{fps} and j_{fps}. Hence, $\theta_g(k, i_{fps}, j_{fps})$ is given by

$$\theta_g(k, i_{fps}, j_{fps}) = \theta_{in}(k, i_{fps}) + [(j_{fps} - 1) - 4]\Delta\theta(k, i_{fps}), \quad (10.20)$$

where

$$\Delta\theta(k, i_{fps}) = \frac{\pi}{4} 2^{-(i_{fps}-1)}. \tag{10.21}$$

In this way, the cost function defines the selection of the optimal (estimated) angle

$$J_{fps} = |0 - v_{g,q}^{+}(k, i_{fps}, j_{fps})|. \tag{10.22}$$

Hence, the angle that allows the minimum value of J_{fps} is got after 64 rounds of the two nested loops. It can be noticed that the FPS does not employ the PI in this loop and it does not need to design gains.

The algorithm assumes that $\theta_{in}(k, i_{fps}) = 0$ rad and $J_{fps} = \infty$, by using Eq. (10.21) in which $i_{fps} = 1$ and $\Delta\theta(k, 1) = \frac{\pi}{4}$ rad in the first round of the outer loop. In this way, using Eq. (10.20), the calculated angles are: $-\pi$, $-\frac{3\pi}{4}$, $-\frac{\pi}{2}$, $-\frac{\pi}{4}$, 0, $\frac{\pi}{4}$, $\frac{\pi}{2}$, $\frac{3\pi}{4}$ rad. Hence, the q-axis component $v_{g,q}^{+}$ is computed using $\theta_g(k, i_{fps}, j_{fps})$ and the angle of the grid that minimizes the cost function is achieved. The mentioned angle will be employed in the second round as $\theta_{in}(k, 2)$, with a precision of the search-based algorithm meliorated by a factor of 2. Again, the grid angular position will be generated using Eq. (10.20), and they are: 0, $\frac{\pi}{8}$, $\frac{\pi}{4}$, $\frac{3\pi}{8}$, $\frac{\pi}{2}$, $\frac{5\pi}{8}$, $\frac{3\pi}{4}$, $\frac{7\pi}{8}$ rad. The algorithm reaches the optimal grid angle when the number of iterations increases, and it has accuracy of $\frac{1}{2} \times \frac{\pi}{4} \times 2^8 = \frac{\pi}{1024} = 0.003$ rad. In this book, the grid angular position is reached until the last interaction ($i_{fps} = 8$).

Fig. 10.5 depicts the diagram representation of the DSOGI-MAF-FPS-PLL in which DSOGI and MAF were employed to obtain the positive sequence of the grid voltage vector in the fundamental frequency even in the presence of harmonic components, and the basic code of the FPS-PLL is presented in Algorithm 10.1.

The grid frequency $\omega_g(k)$ can be computed using Eq. (10.23) as presented in Fig. 10.5, where T is the sampling time,

$$\omega_g(k) = \frac{\theta_g(k) - \theta_g(k-1)}{T}, \tag{10.23}$$

where k and $k - 1$ are the actual and preceding values of the variables during the respective sampling instants. In this way, a filter is employed to eliminate the oscillations/ripples due to the ω_g calculation using (10.23). In this book a second-order Butterworth filter is applied to accomplish this task. The oscillations/ripples can be reduced when the finite angle size is increased, however, it can increase the computational burden.

10.4 Experimental results

The experimental tests are carried out to endorse the performance of the DSOGI-MAF-FPS-PLL. In this way, one permits to operate under voltage sag and nonideal voltages due to harmonics elements and the base voltage is 100 V.

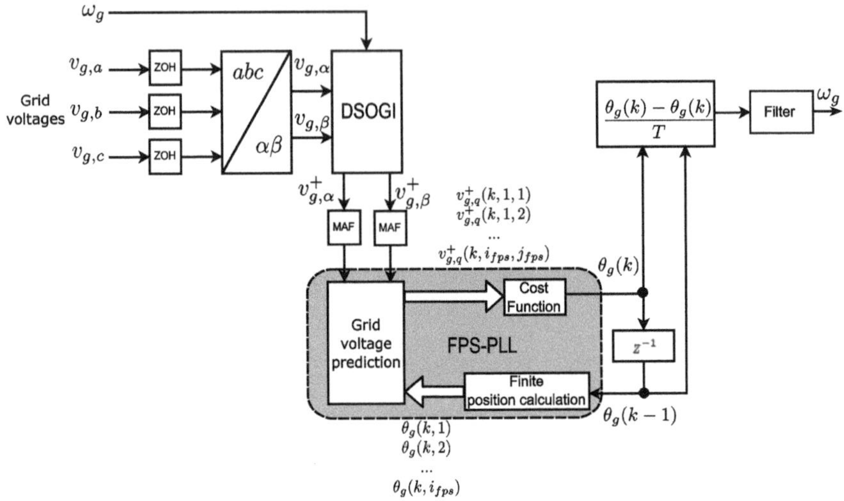

FIGURE 10.5 Diagram of the DSOGI-MAF-FPS-PLL.

Algorithm 10.1 FPS-PLL algorithm

---

while FPS-PLL **do**
  The system variables as voltages are measured
  Positive sequence using DSOGI and MAF
  Set $\theta_{in}(k, i_{fps}) = \theta_g(k-1)$ and $J_{fps-in}(k) = \infty$
  for
  $i_{fps} = 1:8$ **do**
    $\Delta\theta(k, i_{fps}) = \frac{\pi}{4}2^{-(i_{fps}-1)}$
    for
    $j_{fps} = 1:8$ **do**
      $\theta_g(k, i_{fps}, j_{fps}) = \theta_{in}(k, i_{fps}) + [(j_{fps}-1)-4]\Delta\theta(k, i_{fps})$
      Calculation of $v^{+}_{g,q}$ using Eq. (10.19).
      $J_{fps} = |0 - v^{+}_{g,q}(k, i_{fps}, j_{fps})|$
      if $J_{fps}(k, i_{fps}, j_{fps}) < J_{fps-in}(k)$ **then**
        $J_{fps-in}(k) = J_{fps}(k, i_{fps}, j_{fps})$
        $\theta_{g-op}(k) = \theta_g(k, i_{fps}, j_{fps})$ % Store the state with the least cost function
      end if
    end for
  end for
  $\theta_g(k) = \theta_{g-op}(k)$
end while

Fig. 10.6 depicts the performance of the DSOGI-MAF-FPS-PLL operating under unbalanced voltage sag. In this way, two phases decrease their value from the nominal voltage to 53% of nominal voltage. It can be noticed that the FPS-PLL permits operating in this condition, and there is no ripple due to the $2\omega_g$ and the settling time is smaller than 25 ms.

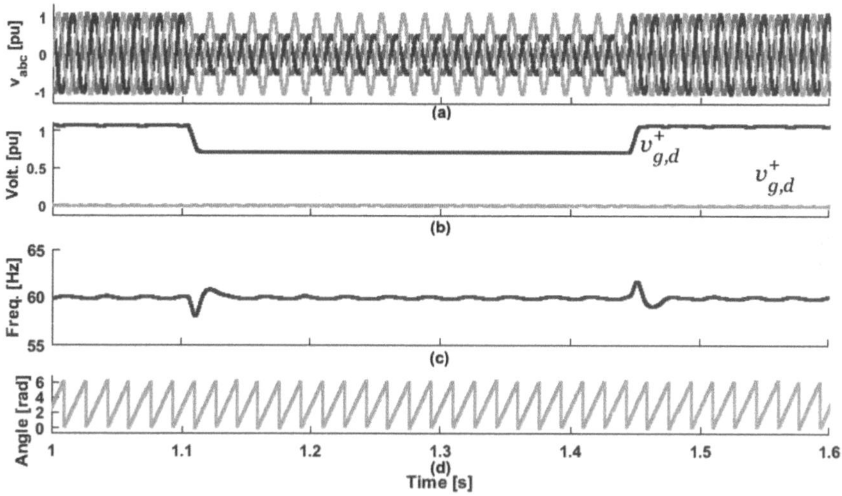

FIGURE 10.6 Experimental result of DSOGI-MAF-FPS-PLL test for unbalanced sag: (a) three-phase voltage, (b) positive sequence of the grid voltages elements in the dq frame, (c) grid frequency, and (d) grid angle.

The performance of the DSOGI-MAF-FPS-PLL operating under balanced voltage sag and the presence of the 5th harmonic component is depicted in Fig. 10.7. In the mentioned test, the 5th harmonic component has 13% of the rated voltage. It can be seen that the DSOGI-MAF-FPS-PLL permits operating in this condition, there is no ripple, and the settling time is smaller than 20 ms again.

The performance of the DSOGI-MAF-FPS-PLL operating under the presence of the harmonics elements is depicted in Fig. 10.8. Table 9.1 introduces the harmonics components values from the nominal voltage as 3.94% for the 5th, 3.15% for the 7th, 2.36% for the 11th, 1.50% for the 13th, 1.10% for the 17th, and 0.70% for the 19th harmonic. It can be seen that the DSOGI-MAF-FPS-PLL permits operating under such conditions and there is no ripple in the frequency or grid angle due to the interference of the harmonic elements.

10.5 Summary

This chapter describes the fundamental concepts of the PLL, and the representation of distorted voltage, DSOGI, MAF, and FPS. In this way, the DSOGI-MAF-FPS-PLL can operate under unbalanced and distorted voltage due to har-

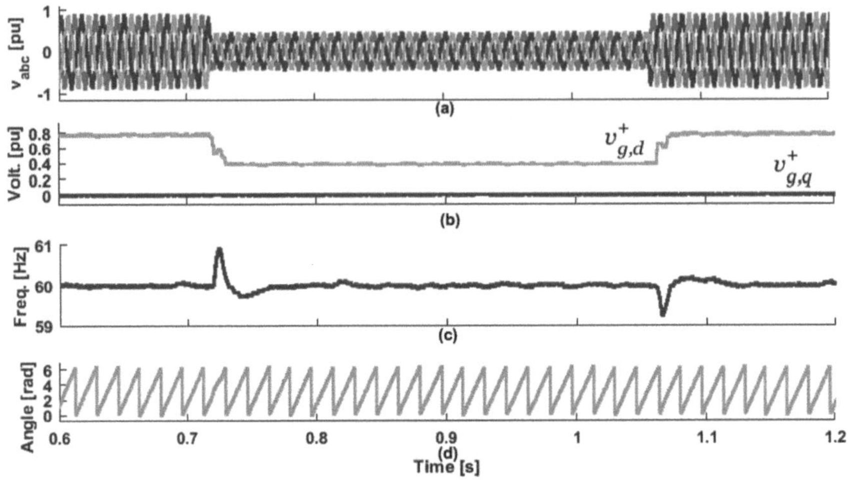

FIGURE 10.7 Experimental result of DSOGI-MAF-FPS-PLL test for voltage sag with the 5th harmonic: (a) three-phase voltage, (b) positive sequence of the grid voltages elements in the *dq* frame, (c) grid frequency, and (d) grid angle.

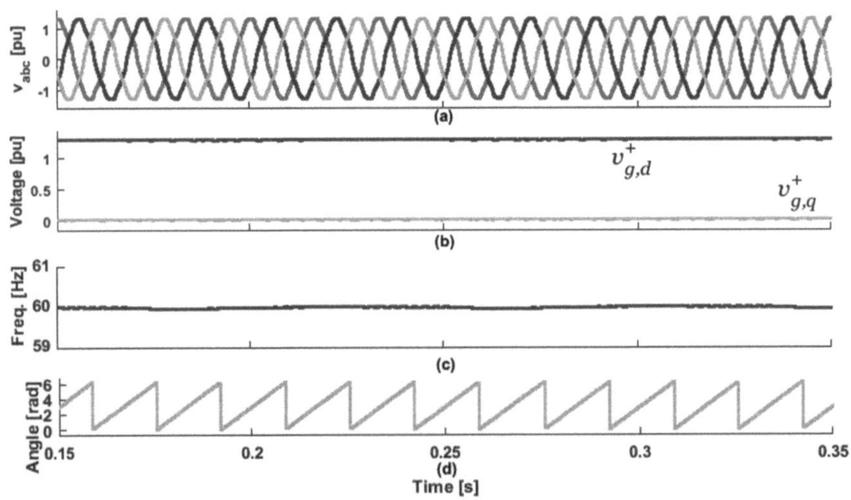

FIGURE 10.8 Experimental result of DSOGI-MAF-FPS-PLL test under distorted voltage: (a) three-phase voltage, (b) positive sequence of the grid voltages elements in the *dq* frame, (c) grid frequency, and (d) grid angle.

monic elements. Hence, DSOGI-MAF-FPS-PLL permits calculating the angle and frequency of the power grid. The experimental results endorse the DSOGI-MAF-FPS-PLL performance under the mentioned conditions.

Chapter 11

Implementation of DFIG MPC-WM and three-phase power CCG MPRC-WM using Simulink/MATLAB®

Alfeu J. Sguarezi Filho, Angelo S. Lunardi, and Eliomar R. Conde D.
Federal University of ABC, Santo André, SP, Brazil

11.1 Introduction

The main idea is to introduce the reader, who is not familiar with MATLAB®/Simulink, to this powerful tool for testing control proposals before experimental testing. To help the reader, a step-by-step procedure was elaborated and is now presented in this chapter not only considering basic instructions like scripts but also complex model and controller implementations. Furthermore, the method presented here is a general way of using the software; in other words, it is not related to a specific version of MATLAB/Simulink.

11.2 Building embedded functions for Park–Clarke transformation

Simulink has blocks for Park and Clarke transformation; and in order to avoid some mistakes or confusion, we present here steps to build your own block for Park and Clarke transformation. All the mathematical equations presented here were already discussed in Chapter 2 and the transformations employed in this book were introduced in Section 3.1.1.2. In this section we produced a Simulink block for Park–Clarke transformation and inverse Park–Clarke transformation.

11.2.1 Park–Clarke transformation

These transformations can be used for both voltage and current.

11.2.1.1 Clarke transformation

The diagram is depicted in Fig. 11.1 and the script code for Clarke transformation (CT) can be described as:

```
CT = [1 -1/2 -1/2; 0 (sqrt(3)/2) (-sqrt(3)/2)];
```

Model Predictive Control for Doubly-Fed Induction Generators and Three-Phase Power Converters
https://doi.org/10.1016/B978-0-32-390964-8.00020-8

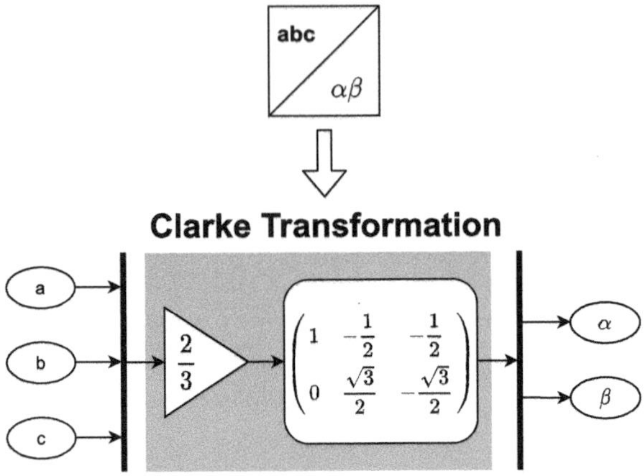

FIGURE 11.1 Clarke transformation (CT).

For matrix multiplication, the reader should use a Simulink block called *Gain* and configure multiplication as *Matrix (K*u)*.

11.2.1.2 Park transformation

Now, Park transformation is the next step to change the system into synchronous frame; so, it is necessary to obtain an angle through PLL. Differently from Clarke transformation, where the reader used the Gain blocks, here we should use the block called *Fcn* which can insert a general expression function, and expand the matrices in two *Fcn* blocks putting *d_Park* at the first block and *q_Park* at the second block. Please, notice that it is necessary to use the *Mux* block and respect the sequence, so the *Fcn* can work correctly as depicted in Figs. 11.2 and 11.3.

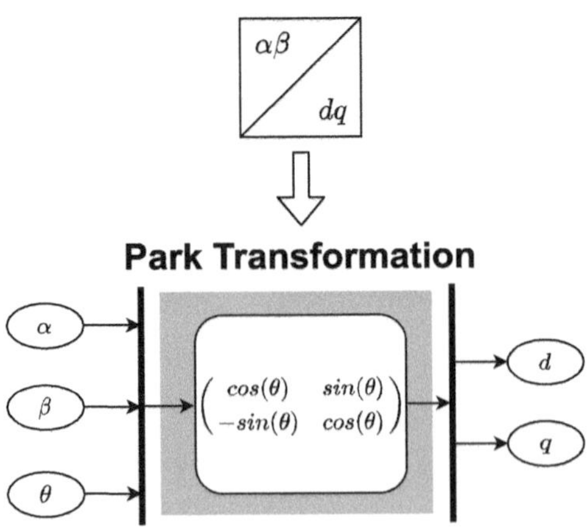

FIGURE 11.2 Park transformation (PT).

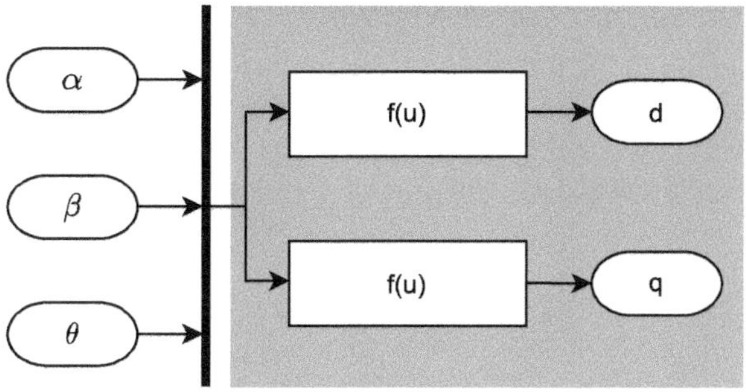

FIGURE 11.3 Park transformation (PT) in Simulink.

```
d_Park = (u(1)*cos(u(3)))+(u(2)*sin(u(3)));
q_Park = -(u(1)*sin(u(3)))+(u(2)*cos(u(3)))
```

11.2.2 Inverse Park–Clarke transformation

The purpose of this inverse transformation is to reverse the effects produced by the original Park–Clarke transformation. Its construction is basically the same as that of the original, but what change are the values and dimensions of the matrices. As the original ones, these can be used for both voltage and current.

11.2.2.1 Inverse Clarke transformation

This is the transformation that reverts a variable from the stationary $\alpha\beta$ frame to the *abc* frame. Fig. 11.4 shows the diagram to construct the transformation.

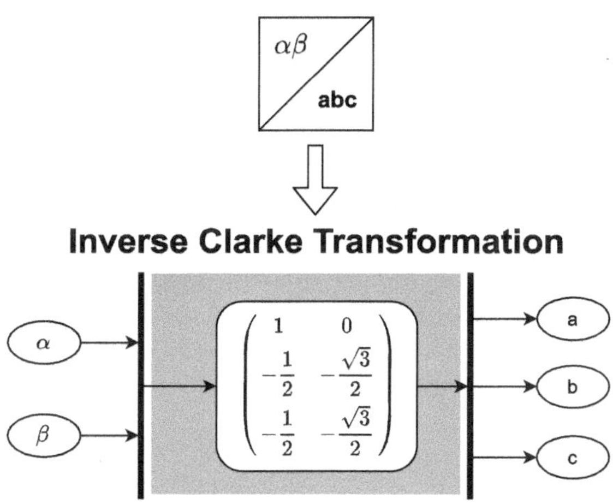

FIGURE 11.4 Inverse Clarke transformation.

11.2.2.2 Inverse Park transformation

Using the same angle of the original Park transformation, this transformation reverts the variables from the synchronous dq frame to the stationary $\alpha\beta$ frame. Fig. 11.5 shows what is being explained.

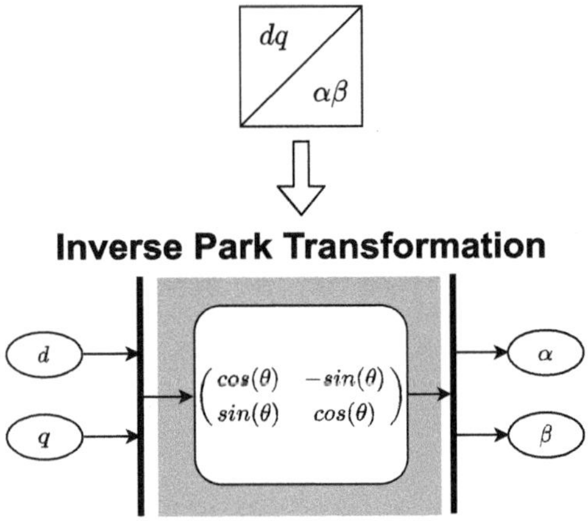

FIGURE 11.5 Inverse Park transformation.

11.2.3 Pulse width modulation

To embed a pulse width modulation into Simulink, there are different ways on how to do it. Two specific techniques will be described in this section.

11.2.3.1 Sinusoidal pulse width modulation (PWM)

This type of modulation is widely applied in power electronics, using a triangular signal and comparing it to a reference signal, such as voltage. The following equation creates the output voltage for the PWM system:

$$j = 1 \text{ if } v_j \geq v_{tri} \text{ with } j = a, b, c,$$

where v_a, v_b, v_c are reference voltages for each phase and v_{tri} is a triangular signal. In Fig. 11.6, the PWM block built in Simulink is presented.

To create this PWM block, it is necessary to use a block to produce the triangular signal; as a suggestion, there is a block called *Repeating* for the triangular signal. The reader should add three more *Relational Operator* blocks and choose the option >= to compare to the triangular signal.

11.2.3.2 Space vector modulation (SVM)

Space vector modulation focuses on finding the time switching for a three-phase inverter based on the voltage to obtain an output sinusoidal voltage.

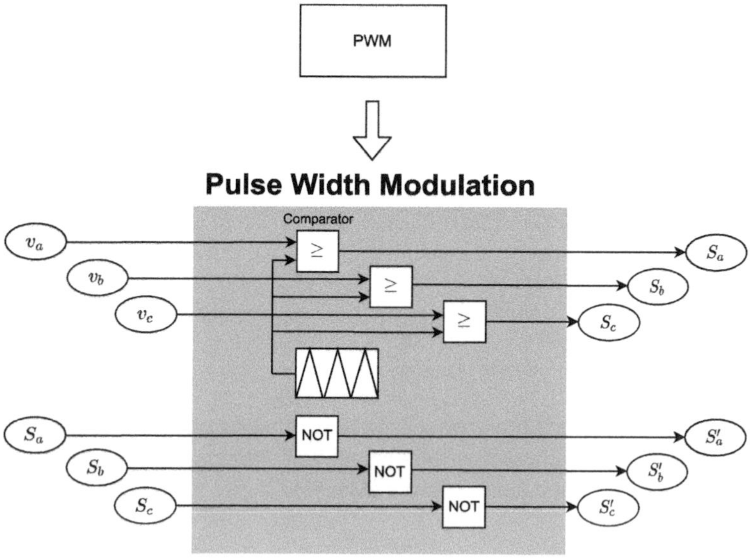

FIGURE 11.6 PWM in Simulink.

The steps to implement the SVM will be presented below. Fig. 11.7 shows the function block used for programming all these steps. The inputs will be the voltage reference, voltage angle, and a triangular signal with a period equal to the sample time and a start at 0 until 100.

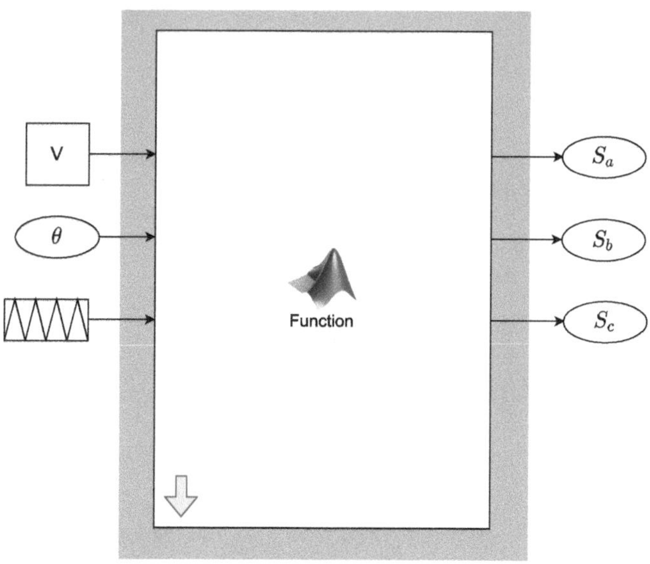

FIGURE 11.7 SVM in Simulink.

1. The first lines of the code declare the control function, output, input, and parameters. The following lines declare the persistent variables across all the simulation with descriptive comments on the right-hand side:

```
function [Sa,Sb,Sc] = SVM(V,theta)
persistent x_opt v2 v1 Sector count r
  %---------- Declaring persistent variables ----------%
if isempty(x_opt), x_opt = 1; end
if isempty(count), count = 0; end
if isempty(r), r = 0; end
if isempty(v1), v1 = 0; end
if isempty(v2), v2 = 60*pi/180; end
if isempty(Sector), Sector = 1; end
```

2. The following code lines are responsible for defining the bus voltage, sample time, states of switches; *a* is the normalized voltage, and the reader should set any values for the first states; here we decided to use zero as the initial value.

```
%-------------------------------------------------------%

Vbus = 300;% Bus Voltage
Tps = 1e-4;% Sample Time

states = [0 0 0;1 0 0;1 1 0;0 1 0;...
          0 1 1;0 0 1;1 0 1;1 1 1]; %States of Switches

a = V/(Vbus); % Normalized Voltage

Sa = 0; %initial value
Sb = 0; %initial value
Sc = 0; %initial value
```

3. The next step is responsible for choosing which sector in the sextant the voltage angle is located in.

```
% Check For Section of Sextant -----------------------%
if theta >= 0*pi/180 && theta < 60*pi/180

    Sector = 1 ;
    v1 = 0*pi/180; % [1 0 0;] states of upper switches
    v2 = 60*pi/180;% [1 1 0;] states of upper switches

elseif theta >= 60*pi/180 && theta < 120*pi/180

    Sector = 2 ;
    v1 = 60*pi/180; % [1 1 0;] states of upper switches
    v2 = 120*pi/180;% [0 1 0;] states of upper switches
```

```
elseif theta >= 120*pi/180 && theta < 180*pi/180

    Sector = 3 ;
    v1 = 120*pi/180;% [0 1 0;] states of upper switches
    v2 = 180*pi/180;% [0 1 1;] states of upper switches

elseif theta >= 180*pi/180 && theta < 240*pi/180

    Sector = 4 ;
    v1 = 180*pi/180;% [0 1 1;] states of upper switches
    v2 = 240*pi/180;% [0 0 1;] states of upper switches

elseif theta >= 240*pi/180 && theta < 300*pi/180

    Sector = 5 ;
    v1 = 240*pi/180;% [0 0 1;] states of upper switches
    v2 = 300*pi/180;% [1 0 1;] states of upper switches

elseif theta >= 300*pi/180 && theta <= 360*pi/180

    Sector = 6 ;
    v1 = 300*pi/180;% [1 0 1;] states of upper switches
    v2 = 360*pi/180;% [1 0 0;] states of upper switches

end
%--------------------------------------------------%
```

4. Now in order to define the switching times of the vector components, the following code is responsible for calculating each time:

```
% Switching times of vector components -----------------%
Ta = Tps*a*(sin(v2 - theta)/sin(v2 - v1));

Tb = Tps*a*(sin(theta - v1)/sin(v2 - v1));

T0 = Tps - Ta - Tb;

%--------------------------------------------------%
```

5. Finally, the code to choose the switches' states based on T_a, T_b, and T_0 is presented below. Notice that if the conditional loop `If` is present, it will be necessary to rescale T_a, T_b, and T_0 because the triangular signal is from 0 to 100.

```
% Compare of Triangular Signal and Select Switches -------%

switch Sector
```

```
case 1

    if count <= (T0/(2*Tps))*100
        x_opt = 1;
        Sa = states(x_opt,1);
        Sb = states(x_opt,2);
        Sc = states(x_opt,3);

    elseif count <= ((T0/2+Ta/1)/Tps)*100
        x_opt = 2;
        Sa = states(x_opt,1);
        Sb = states(x_opt,2);
        Sc = states(x_opt,3);

    elseif count <= ((T0/2+Ta/1+Tb/1)/Tps)*100
        x_opt = 3;
        Sa = states(x_opt,1);
        Sb = states(x_opt,2);
        Sc = states(x_opt,3);

    else
        x_opt = 8;
        Sa = states(x_opt,1);
        Sb = states(x_opt,2);
        Sc = states(x_opt,3);

    end

case 2

    if count <= (T0/(2*Tps))*100
        x_opt = 1;
        Sa = states(x_opt,1);
        Sb = states(x_opt,2);
        Sc = states(x_opt,3);

    elseif count <= ((T0/2+Ta/1)/Tps)*100
        x_opt = 3;
        Sa = states(x_opt,1);
        Sb = states(x_opt,2);
        Sc = states(x_opt,3);

    elseif count <= ((T0/2+Ta/1+Tb/1)/Tps)*100
        x_opt = 4;
        Sa = states(x_opt,1);
        Sb = states(x_opt,2);
        Sc = states(x_opt,3);
```

```
    else
        x_opt = 8;
        Sa = states(x_opt,1);
        Sb = states(x_opt,2);
        Sc = states(x_opt,3);

    end

case 3

    if count <= (T0/(2*Tps))*100
        x_opt = 1;
        Sa = states(x_opt,1);
        Sb = states(x_opt,2);
        Sc = states(x_opt,3);

    elseif count <= ((T0/2+Ta/1)/Tps)*100
        x_opt = 4;
        Sa = states(x_opt,1);
        Sb = states(x_opt,2);
        Sc = states(x_opt,3);

    elseif count <= ((T0/2+Ta/1+Tb/1)/Tps)*100
        x_opt = 5;
        Sa = states(x_opt,1);
        Sb = states(x_opt,2);
        Sc = states(x_opt,3);

    else
        x_opt = 8;
        Sa = states(x_opt,1);
        Sb = states(x_opt,2);
        Sc = states(x_opt,3);

    end

 case  4

    if count <= (T0/(2*Tps))*100
        x_opt = 1;
        Sa = states(x_opt,1);
        Sb = states(x_opt,2);
        Sc = states(x_opt,3);

    elseif count <= ((T0/2+Ta/1)/Tps)*100
        x_opt = 5;
        Sa = states(x_opt,1);
        Sb = states(x_opt,2);
```

```
            Sc = states(x_opt,3);

        elseif count <= ((T0/2+Ta/1+Tb/1)/Tps)*100
            x_opt = 6;
            Sa = states(x_opt,1);
            Sb = states(x_opt,2);
            Sc = states(x_opt,3);

        else
            x_opt = 8;
            Sa = states(x_opt,1);
            Sb = states(x_opt,2);
            Sc = states(x_opt,3);

        end

    case  5

        if count <= (T0/(2*Tps))*100
            x_opt = 1;
            Sa = states(x_opt,1);
            Sb = states(x_opt,2);
            Sc = states(x_opt,3);

        elseif count <= ((T0/2+Ta/1)/Tps)*100
            x_opt = 6;
            Sa = states(x_opt,1);
            Sb = states(x_opt,2);
            Sc = states(x_opt,3);

        elseif count <= ((T0/2+Ta/1+Tb/1)/Tps)*100
            x_opt = 7;
            Sa = states(x_opt,1);
            Sb = states(x_opt,2);
            Sc = states(x_opt,3);

        else
            x_opt = 8;
            Sa = states(x_opt,1);
            Sb = states(x_opt,2);
            Sc = states(x_opt,3);

        end


    case  6

        if count <= (T0/(2*Tps))*100
```

```
        x_opt = 1;
        Sa = states(x_opt,1);
        Sb = states(x_opt,2);
        Sc = states(x_opt,3);

    elseif count <= ((T0/2+Ta/1)/Tps)*100
        x_opt = 7;
        Sa = states(x_opt,1);
        Sb = states(x_opt,2);
        Sc = states(x_opt,3);

    elseif count <= ((T0/2+Ta/1+Tb/1)/Tps)*100
        x_opt = 2;
        Sa = states(x_opt,1);
        Sb = states(x_opt,2);
        Sc = states(x_opt,3);

    else
        x_opt = 8;
        Sa = states(x_opt,1);
        Sb = states(x_opt,2);
        Sc = states(x_opt,3);

    end

end
%----------------------------------------------------------%
```

Notice that the function blocks' outputs are the only three signals, which then will be used for the inverter top switches. In order to complete the six switches, the reader should add a *NOT* block for the inverter bottom switches.

11.2.3.3 Three-phase power converter model

The voltage that will be applied on the grid after PWM needs to pass through a three-phase converter; there is a model in which the signals from PWM become three-phase voltage; in other words, each voltage is referred to the neutral point. This block is presented in Fig. 11.8. This configuration is interesting to the rotor or grid side converter and, in this case, the switches have ideal actions. In this way, the state of the top switches and the voltage at DC link (V_{bus}) are enough to compute the three-phase voltages.

11.2.3.4 Three-phase grid voltages model

Fig. 11.9 introduces the representation of the three-phase grid voltages' model. In this way, each voltage is referred to a neutral point operating with frequency ω, angle ωt, and the peak value of the voltage being $|v|$. This configuration is interesting for the DFIG and CCG models.

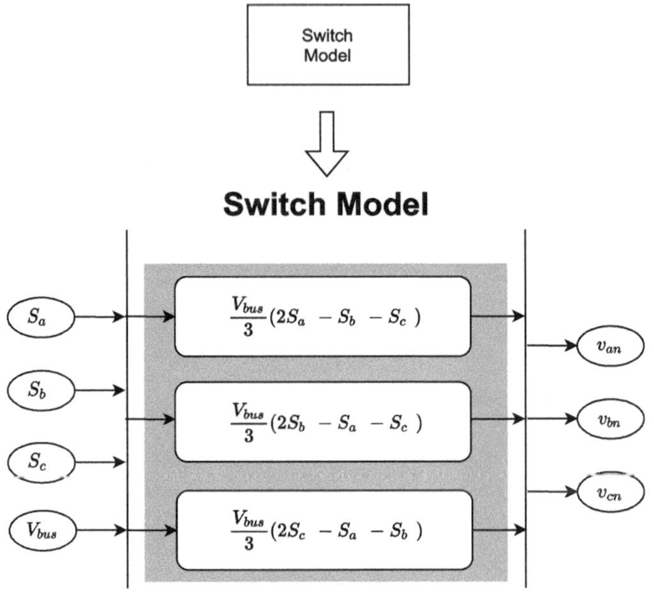

FIGURE 11.8 Model for three-phase power converter in Simulink.

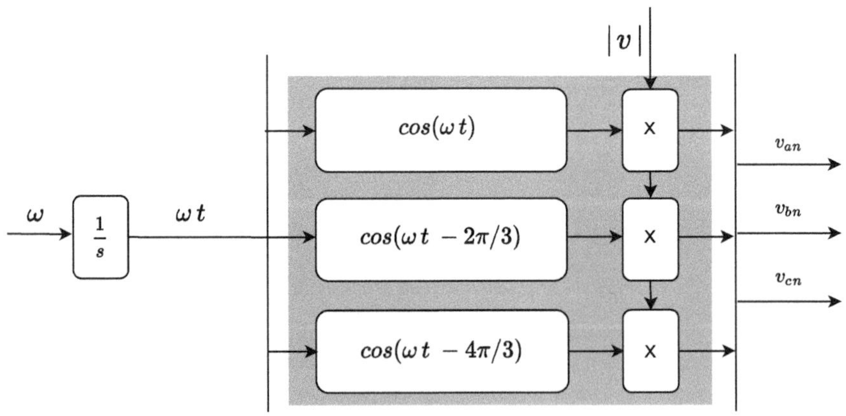

FIGURE 11.9 Three-phase grid voltages model.

Another possibility is to employ the *PLL (3ph)* of the MATLAB/Simulink to compute the angle.

11.3 Building simulation model for DFIG

As stated in the introduction to Chapter 2, this section intends to present the step-by-step construction of the simulation model for the DFIG on MAT-

LAB/Simulink. For this, it is necessary to remember some expressions of the DFIG mathematical model. So, as it is well-known, the expressions for the voltages and fluxes from the stator reference frame and in space vector notation ($\alpha\beta$ frame) are given by:

$$\vec{v}_{s,\alpha\beta} = R_s \vec{i}_{s,\alpha\beta} + \frac{d\vec{\lambda}_{s,\alpha\beta}}{dt}, \tag{11.1}$$

$$\vec{v}_{r,\alpha\beta} = R_r \vec{i}_{r,\alpha\beta} + \frac{d\vec{\lambda}_{r,\alpha\beta}}{dt} - j\omega_m \vec{\lambda}_{r,\alpha\beta}, \tag{11.2}$$

$$\vec{\lambda}_{s,\alpha\beta} = L_s \vec{i}_{s,\alpha\beta} + L_m \vec{i}_{r,\alpha\beta}, \tag{11.3}$$

$$\vec{\lambda}_{r,\alpha\beta} = L_m \vec{i}_{s,\alpha\beta} + L_r \vec{i}_{r,\alpha\beta}. \tag{11.4}$$

The electromagnetic torque is given by

$$T_{em} = \frac{3}{2} L_m NP Im\{\vec{i}_{s,\alpha\beta}.\vec{i}^{\,*}_{r,\alpha\beta}\} = \frac{2}{3} NP L_m (i_{\alpha,r} i_{\beta,s} - i_{\alpha,s} i_{\beta,r}), \tag{11.5}$$

with NP as the pair of poles of the machine.

Now, if Eqs. (11.1)–(11.4) are rearranged taking the currents as state-space magnitudes, the DFIG model will be given as follows [3]:

$$\begin{aligned} \frac{d}{dt}\begin{bmatrix} \vec{i}_{s,\alpha\beta} \\ \vec{i}_{r,\alpha\beta} \end{bmatrix} &= \left(\frac{1}{\sigma L_s L_r}\right) A_{sim}. \begin{bmatrix} \vec{i}_{s,\alpha\beta} \\ \vec{i}_{s,\alpha\beta} \end{bmatrix} \\ &+ \left(\frac{1}{\sigma L_s L_r}\right) \begin{bmatrix} L_r & -L_m \\ -L_m & L_s \end{bmatrix} . \begin{bmatrix} \vec{v}_{s,\alpha\beta} \\ \vec{v}_{r,\alpha\beta} \end{bmatrix}, \end{aligned} \tag{11.6}$$

with A_{sim} being

$$A_{sim} = \begin{bmatrix} -R_s L_r - j\omega_m {L_m}^2 & R_r L_m - j\omega_m L_m L_r \\ R_s L_m + j\omega_m L_m L_s & -R_r L_s + j\omega_m L_r L_s \end{bmatrix}. \tag{11.7}$$

Remembering that $\sigma = 1 - {L_m}^2 / L_s L_r$. So, if the previous model is expanded into the $\alpha\beta$ components then

$$\begin{aligned} \frac{d}{dt}\begin{bmatrix} i_{s,\alpha} \\ i_{s,\beta} \\ i_{r,\alpha} \\ i_{r,\beta} \end{bmatrix} &= \left(\frac{1}{\sigma L_s L_r}\right) A_{\alpha\beta}. \begin{bmatrix} i_{s,\alpha} \\ i_{s,\beta} \\ i_{r,\alpha} \\ i_{r,\beta} \end{bmatrix} \\ &+ \left(\frac{1}{\sigma L_s L_r}\right) \begin{bmatrix} L_r & 0 & -L_m & 0 \\ 0 & L_r & 0 & -L_m \\ -L_m & 0 & L_s & 0 \\ 0 & -L_m & 0 & L_s \end{bmatrix} . \begin{bmatrix} v_{s,\alpha} \\ v_{s,\beta} \\ v_{r,\alpha} \\ v_{r,\beta} \end{bmatrix}. \end{aligned} \tag{11.8}$$

Then $A_{\alpha\beta}$ is

$$A_{\alpha\beta} = \begin{bmatrix} -R_s L_r & \omega_m {L_m}^2 & R_r L_m & \omega_m L_m L_r \\ -\omega_m {L_m}^2 & -R_s L_r & -\omega_m L_m L_r & R_r L_m \\ R_s L_m & \omega_m L_s L_m & -R_r L_s & -\omega_m L_r L_s \\ \omega_m L_s L_m & R_s L_m & \omega_m L_r L_s & -R_r L_s \end{bmatrix}. \quad (11.9)$$

Thus, the construction of the simulation model will be divided into three main blocks, namely currents', fluxes', and torque blocks. The union of these three blocks and the use of the Clarke and Park transformations make the complete DFIG simulation model.

The currents' block corresponds to the state model in (11.8), where it receives as inputs the voltages and currents of rotor and stator in the stationary frame ($\alpha\beta$). This block can be built as detailed Fig. 11.10.

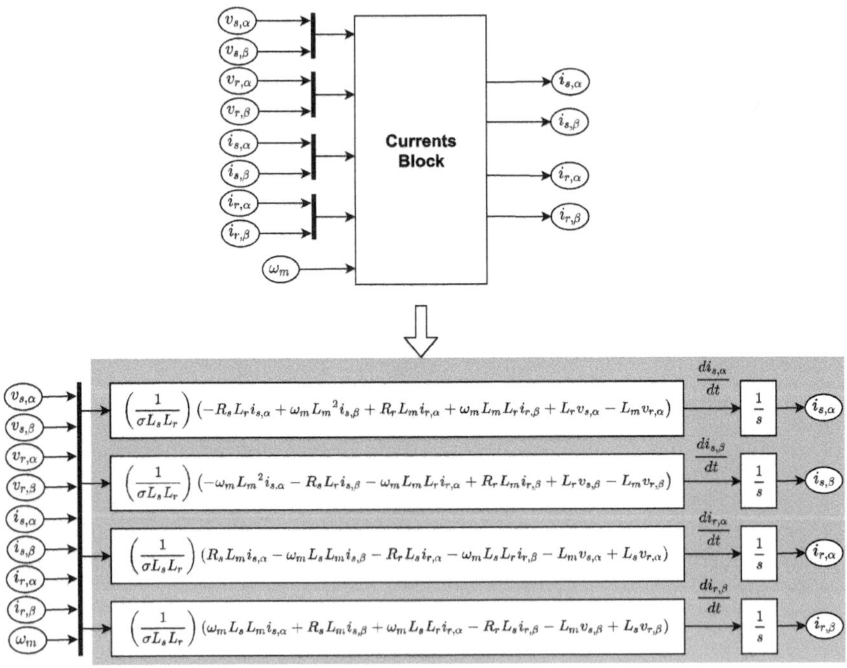

FIGURE 11.10 Currents' block of DFIG simulation model.

Secondly, the fluxes' block is built by means of Eqs. (11.3) and (11.4) using the stator and rotor $\alpha\beta$ currents, which is illustrated in Fig. 11.11.

Finally, the third and last block, i.e., the torque block, is built using the expression (11.5), but this block has a peculiarity – in this DFIG model the rotor speed is an input, it happens because it is easier to the study system speed when it is an input and not the load torque. The mechanical model of the system is still

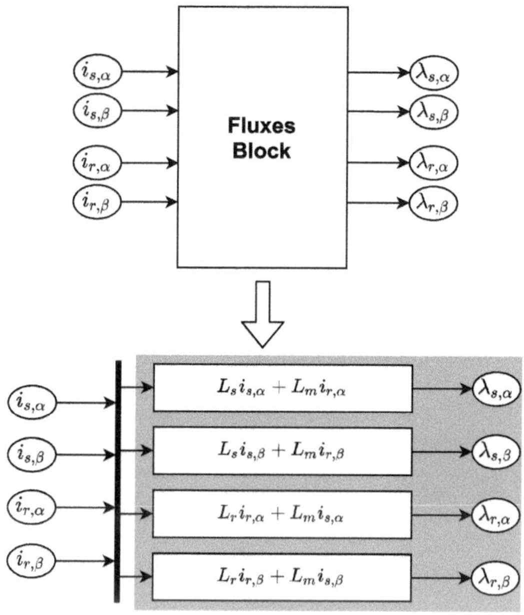

FIGURE 11.11 Fluxes' block of DFIG simulation model.

governed by an equation as $T_{em} - T_{load} = J\frac{d\Omega_m}{dt}$, where Ω_m is the mechanical speed of the rotor in revolutions per minute (rpm), but the difference is that the speed is an input and, if you want to obtain the load torque, you just have to obtain it from the equation. Thus, the torque block is presented in Fig. 11.12.

Once the three main blocks are built, they are put together to form the complete model of the DFIG. This model assumes that the input voltages are given in the real *abc* phase, so it is necessary to use Clarke and Park transformations to carry them to the stator stationary frame ($\alpha\beta$). Transformations are also used after obtaining the stator and rotor currents in the $\alpha\beta$ frame to carry them to the *abc* frame, as it is shown in Fig. 11.13.

Note that in Fig. 11.13, besides the Clarke and Park transformations, there is a factor u, which corresponds to the stator/rotor turns ratio. The transformations employed in this book were introduced in Section 3.1.1.2. With this simulation model, the reader is able to manipulate and study all the output variables of the system or even its internal variables (e.g., $\alpha\beta$ currents). The system itself can be summarized by listing inputs, outputs, and parameters:

Inputs. Stator and rotor voltages (abc) and the rotor mechanical speed (rpm);

Outputs. Stator and rotor currents (abc), stator and rotor fluxes ($\alpha\beta$), electromagnetic torque T_{em}, speed (rad/s) and the rotor angle.

Parameters. R_s, R_r, L_s, L_r, L_m, p, and u.

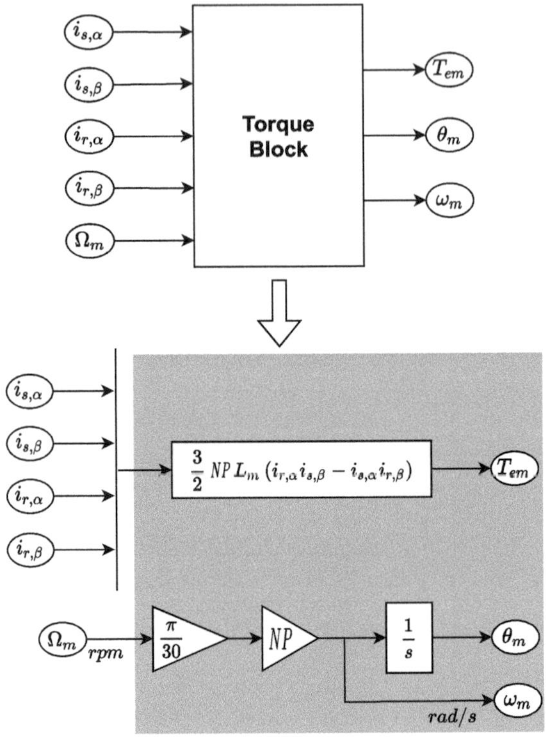

FIGURE 11.12 Torque block of DFIG simulation model.

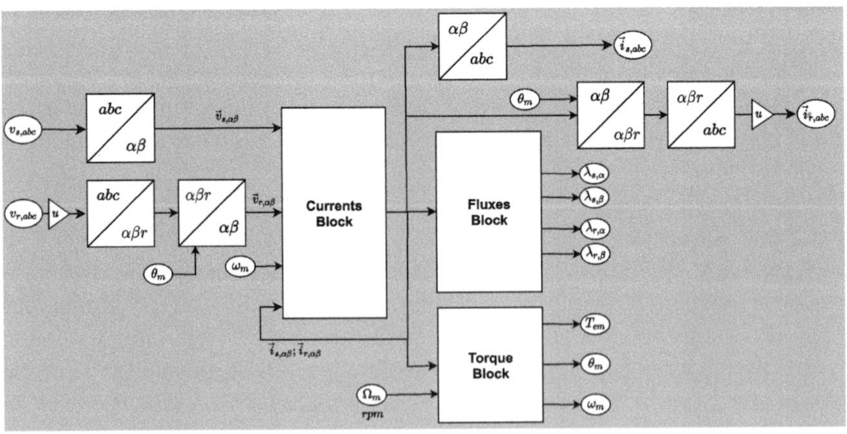

FIGURE 11.13 DFIG system simulation block.

To set the system parameters, the reader has two options when using MATLAB/Simulink. The first is to create an ".m" file setting all the necessary parameters, while the second is the creation of masks on every block, using the mask editor of Simulink. This document uses the first option, which is the creation of the ".m" file that sets all the parameters for the model to run. The following is an illustration of an example sketch:

```
%Setup file for DFIG parameters

Rs = 1 % Ohms
Rr = 3.1322 % Ohms
Lm = 0.1917 % H
Ls = 0.2010 % H
Lr = 0.2010 % H
NP = 2 % Pair of poles
u = 1 % Turns ratio

sigma = 1 - ((Lm^2)/(Ls*Lr)) % sigma parameter
```

The last step before running the system model is to run this test sketch in order to have all the parameters set.

11.3.1 Building simulation model for DFIG using MPC-WM

Model predictive control is the strategy described in Section 7.1, where the controller uses the mathematical model of the system to optimize and determine the future control actions over it. Thus, this subsection is dedicated to introduce the reader to a way of building this controller on Simulink software.

Bringing back the output predictive equation from Section 7.1, it is given by

$$YY_{ir} = AA_{ir}x_{ir}(k) + BB_{ir}VV_{ir} + GG_{ir}DD_{ir}. \tag{11.10}$$

And the output minimized equation that contains the control actions is given by

$$\begin{aligned} VV_{ir} = {} & (BB_{ir}^T W_{yy,ir} BB_{ir} + W_{uu,ir})^{-1} \cdot BB_{ir}^T W_{yy,ir} \\ & \cdot (YY_{ref,ir} - AA_{ir}x_x - GG_{ir}DD_{ir}). \end{aligned} \tag{11.11}$$

The purpose of the controller is to solve this last Eq. (11.11). So in order to achieve this, all the matrices inside of it should be built.

The system that is going to be controlled must be represented in the state space model, like the expression below

$$\underbrace{\begin{bmatrix} i_{r,d}(k+1) \\ i_{r,q}(k+1) \end{bmatrix}}_{x_{ir}(k+1)} = \underbrace{\begin{bmatrix} 1 - \dfrac{R_r T}{\sigma L_r} & \omega_{sl} T \\ -\omega_{sl} T & 1 - \dfrac{R_r T}{\sigma L_r} \end{bmatrix}}_{A_{ad,ir}} \underbrace{\begin{bmatrix} i_{r,d}(k) \\ i_{r,q}(k) \end{bmatrix}}_{x_{ir}(k)}$$

$$+\underbrace{\begin{bmatrix} \frac{T}{\sigma L_r} & 0 \\ 0 & \frac{T}{\sigma L_r} \end{bmatrix}}_{B_{bd,ir}} \underbrace{\begin{bmatrix} v_{r,d}(k) \\ v_{r,q}(k) \end{bmatrix}}_{u_{u,ir}(k)} + \underbrace{\begin{bmatrix} 1 & 0 \\ 0 & 1 \end{bmatrix}}_{G_{gd,ir}} \underbrace{\begin{bmatrix} 0 \\ -\frac{\omega_{sl} L_m T}{\sigma L_r L_s} \lambda_s(k) \end{bmatrix}}_{d_{d,ir}(k)}. \tag{11.12}$$

Taking all this into account, Fig. 11.14 shows an external diagram of what the controller will be, showing it with inputs and outputs. This controller will be built using the "MATLAB function" block in Simulink, which allows writing a function that is then executed within the Simulink environment. Note that this diagram does not contain zeroth order hold (ZOH) at the inputs; however, it is highly recommended to use ZOH at the inputs.

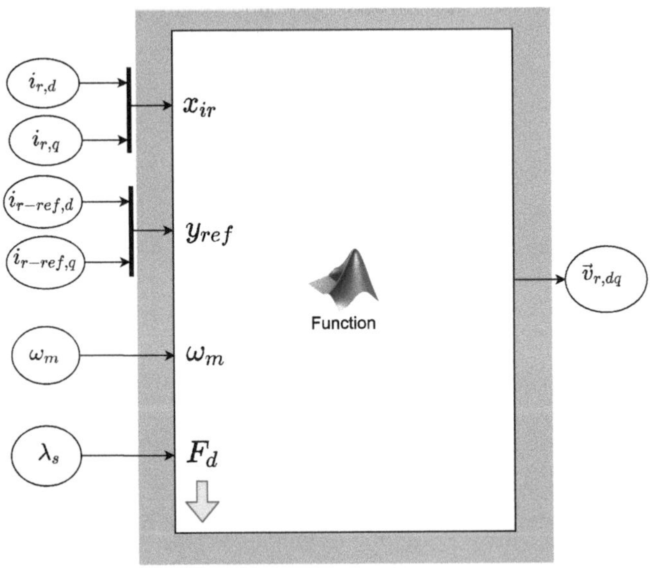

FIGURE 11.14 Model predictive control block, MATLAB/Simulink implementation.

This block can also be summarized by describing inputs, outputs, and parameters:

Inputs. Rotor currents and rotor current references (dq), rotor angular speed (rad/s) and the magnitude of the stator flux λ_s.

Outputs. Rotor voltage in the dq frame.

Parameters. R_r, L_s, L_r, L_m, σ, ω_s, and T_s.

After choosing the "MATLAB function" block, setting the input, output, and parameters of the block, it is time to write the implementation code of the controller. The code will be explained step-by-step to facilitate understanding:

1. The first lines of the code declare the control function, output, input, and parameters. The following lines declare the persistent variables across all the simulation with descriptive comments on the right-hand side:

```
function vdq = Pred(x_ir, wm, y_ref_ir, Fd,...
... Rr, Lr, ws, sigma, Lm, Ls, Ts)
  %---------- Declaring persistent variables ----------%
    persistent YY_ref_ir; % References matrix
    persistent dd_ir; % Perturbation vector
    persistent Wsl; % Slip speed
    persistent F_d; % Stator flux magnitude
    persistent B_bd_ir; % input matrix Bd
    persistent AA_ir; % Predictive Matrix
    persistent BB_ir; % Control matrix
    persistent GG_ir;  % Perturbation matrix
```

2. The second step is the declaration and initialization of normal variables, followed by initialization of persistent variables:

```
  %--------- Declarations and Initializations ---------%
    nInputs = 2; % Inputs of the state space model
    numStates = size(x_ir,1);
    % Number of states of the state space model
    numOutputs = size(y_ref_ir,1);
    % Number of outputs of the state space model
    numDisturbances = 2;
    % size of the disturbance matrix of the state space model
    wsl = ws - wm; % Slip speed

    %-- Predictive parameters --%
    predHorizon = 3;
    ctrlHorizon = 2;

     %-- Initialization of persistens --%
   if isempty(YY_ref_ir);
        YY_ref_ir = zeros( predHorizon * numOutputs,1);
    end
   if isempty(dd_ir);
       dd_ir = zeros(numDisturbances, 1);
    end
    if isempty(Wsl);
        Wsl = ws;
    end
    if isempty(F_d);
        F_d = Fd; %Assigning input stator flow to persistent flow
    end
```

3. Third, it is now time to calculate matrices, they are computed only if some change occurs in the state space model (if the slip speed ω_{sl} changes) or if it is

the first calculation of the predictive matrix from Eq. (11.10). The predictive control matrices are calculated with dedicated functions that will be explained later.

```
%----------- Computation of the matrices ------------%
  if ( (Wsl ~= wsl) || (isempty(Fx)) )
      Wsl = wsl; % assign a new value to the persistent slip speed

    %----------- Creating State Matrices ------------%
    %-- Filling matrix A_ad_ir --%
      A_ad11 = (1-(Rr*Ts)/(sigma*Lr));
      A_ad_ir = [A_ad11, wsl*Ts; -wsl*Ts, A_ad11];
      C_cd_ir = eye(numOutputs);

    %-- Filling persistent matrix B_bd_ir --%
     if isempty(B_bd_ir);
          B_bd11 = Ts/(sigma*Lr);
          B_bd_ir = [B_bd11, 0; 0, B_bd11];
      end

      F_d = Fd; %assigning persistent value to flux
      % Creating the disturbance vector
      dd_ir = [ 0; (-wsl*Lm*Ts)*F_d/(sigma*Ls*Lr)];

    %------ Generating the Predictive control matrices ------%
      AA_ir = AA_ir_Matrix(A_ad_ir, predHorizon, C_cd_ir);
      BB_ir = BB_ir_Matrix(A_ad_ir, B_bd_ir,...
      ...predHorizon, ctrlHorizon, C_cd_ir);
     GG_ir = GG_ir_Matrix(A_ad_ir, C_cd_ir, predHorizon);
  end
```

4. The fourth step is the creation of the weighting matrices using the size of the control matrix "Phi" (Φ). Also a conditional block is implemented to see if the stator flux changed and thus the disturbance vector got modified.

```
%------ Generating the weighting matrixes ------%
  [m,n] = size(Phi);
  W_yy_ir = 2e10*eye(m);
  W_uu_ir = 3e5*eye(n);

%----------- Perturbation vector -----------%
  if ( (F_d ~= Fd));
      F_d = Fd; %assigning persistent value to flux
   dd_ir = [ 0; (-wsl*Lm*Ts)*F_d/(sigma*Ls*Lr)];
  end
```

5. Now, the references' vector is built, repeating the current reference input (y_{ref}) with the prediction horizon, the same is done with the disturbance vector

(ω_{dk}) and the disturbance ω_d. After this, the most important step must be undertaken, which is the computation of the output minimized equation, Eq. (11.11).

```
% References and disturbance vectors
  YY_ref_ir = repmat(y_ref_ir, predHorizon, 1);
  % Creating references vector
  DD_ir = repmat(dd_ir, predHorizon, 1); %creating disturbances vector

% minimized cost function
  VV_ir = ((BB_ir'*W_yy_ir*BB_ir + W_uu_ir)^-1)*BB_ir'...
          *W_yy_ir*(YY_ref_ir -AA_ir*x_ir -GG_ir*DD_ir);
```

In the third step, there are three functions that help in the construction of the matrices for the predictive control. The content of these matrices and how they are constructed are explained in Section 7.1 together with the controller, in general.

The three functions are:

```
function [AA_ir] = AA_ir_Matrix(A, predHorizon, C)
    [row, col] = size(A);
    coder.varsize('AA_ir', [row*(predHorizon) col]);
    AA_ir = C*A;
    Aaux = C*A;
    for i = 2: predHorizon
        Aaux = Aaux*A;
        AA_ir = [ AA_ir; Aaux];
    end
end

function [BB_ir] = BB_ir_Matrix(A, B, predHorizon, ctrlHorizon, C)
    [aux, colSize] = size(B);
    [rowSize, aux] = size(C);
    zeroA = zeros(rowSize, colSize*(ctrlHorizon-1));
    BB_irColumns = colSize*ctrlHorizon;
    coder.varsize('BB_ir', [rowSize*(predHorizon) BB_irColumns]);
    BB_ir = [ C*B, zeroA];
    Aux = BB_ir;
    for i = 1: (predHorizon - 1)
        Aux = [C*(A^(i))*B, Aux];
        Aux = Aux(:,1:BB_irColumns);
        BB_ir = [ BB_ir; Aux];
    end
end

function [GG_ir] = GG_ir_Matrix(A, C, predHorizon)
    [rowSize, colSize] = size(C);
    zeroA = zeros(rowSize, colSize*(predHorizon-1));
    DColumns = colSize*predHorizon;
```

```
    coder.varsize('D', [rowSize*(predHorizon) DColumns]);
    GG_ir = [ C, zeroA];
    Aux = GG_ir;
    for i = 1: (predHorizon - 1)
        Aux = [ C*(A^(i)), Aux];
        Aux = Aux(:, 1:DColumns);
        GG_ir = [ GG_ir; Aux];
    end
end
```

It is important to highlight that this code can be also implemented experimentally on some DSP or other experimental board, doing the corresponding modifications and simplifications needed. Finally, the authors encourage the reader to implement this controller on Simulink using the DFIG model already presented in the previous section.

11.3.2 Building simulation model for DFIG using MPRC-WM

In this section, the controller that will be implemented is the MPRC for DFIG which is explained in Section 9.3. This control strategy is very similar to that explained in the last section, it uses the same state space model but differs a little in the equations; here it is assumed that the distortion is being eliminated by a generator polynomial that is introduced (for more information, see Section 9.3). As the last subsection, this one shows how to implement this controller using Simulink software.

Recall the predictive equation for this controller, namely

$$I_{pr,ir} = A_{pr,ir} x_{ee,ir}(k_i) + B_{pr,ir} V_{pr,ir}. \tag{11.13}$$

The minimization of the corresponding cost function yields

$$V_{pr,ir} = -(B_{pr,ir}^T \overline{Q_{i,ir}} B_{pr,ir} + \overline{R_{v,ir}})^{-1} B_{pr,ir}^T \overline{Q_{i,ir}} A_{pr,ir} x_{ee,ir}(k_i). \tag{11.14}$$

As the system to be controlled is the same as that of the MPC implementation, the state space model is the same as in the last section — the only difference is that considering this controller, an expanded state space model must be built and, after that, the predictive matrices have to be built with this expanded state space.

Again, as per Section 9.3, the mentioned expanded state space model has the following form:

$$\begin{aligned} x_{ee,ir}(k+1) &= A_{ee,ir} x_{ee,ir}(k) + B_{ee,ir} u_{xx,ir}(k), \\ y_{y,ir}(k) &= C_{ee,ir} x_{ee,ir}. \end{aligned} \tag{11.15}$$

This new expanded model has the particularity that it does not have a disturbance vector, and that is the reason why Eq. (11.13) has that form without any disturbance. As the MPC implementation, this predictive repetitive block is built using the "MATLAB function" from Simulink, and Fig. 11.15 shows a general scheme of the controller.

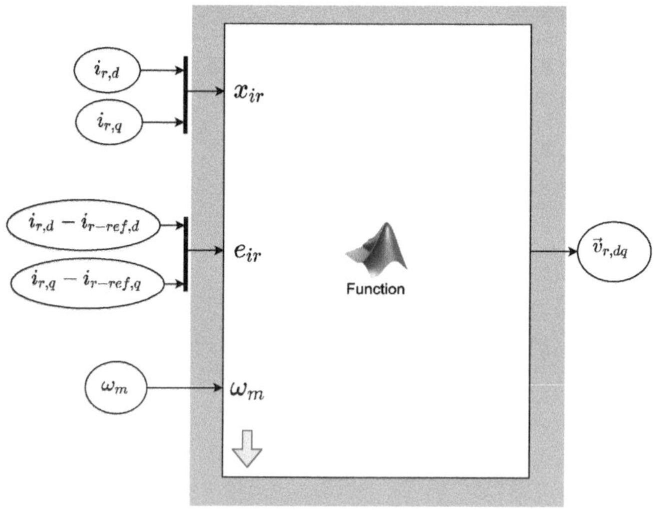

FIGURE 11.15 Predictive Repetitive control block, MATLAB/Simulink implementation.

In this case it is also recommended to use ZOH blocks at the inputs.

The inputs, outputs, and parameters of this block are:

Inputs. Rotor currents ($\vec{i}_{r,dq}$), error between rotor currents and its references ($\vec{i}_{r,dq} - \vec{i}_{r-ref,dq}$) and rotor angular speed ω_m (rad/s);

Outputs. Rotor voltage in the dq frame $\vec{v}_{r,dq}$;

Parameters. R_r, L_s, L_r, L_m, σ, ω_s, T_s, and $D(z)$.

Note that compared to the standard MPC one additional parameter of this controller is the polynomial $D(z)$, which in this case is specified as $1 - z^{-1}$. The code contained inside this MATLAB function block is explained step-by-step in the following paragraphs:

1. The first lines of the code declare the control `PredRep` function with output, input, and parameters. The following lines declare the persistent variables across all the simulation with descriptive comments on the right-hand side:

```
function vdq = PredRep(x_ir, wm, e_ir, Rr, Lr, ws, sigma, D, Ts)
  %---------- Declaring persistent variables ----------%
    persistent X_ir; % current and past state vector
    persistent E_ir; % errors matrix
    persistent U_ir; % Past control signals
    persistent Wsl; % Slip speed
    persistent B_bd_ir; % input matrix Bd from state space
          persistent A_pr_ir; % Predictive Matrix
          persistent B_pr_ir; % Control matrix
```

2. Initialization of miscellaneous and persistent variables:

```
%---------- Declarations and Initializations ----------%
  numCoeff = size(D,2); % Number of coefficients of the polynomial
  nInputs = 2; %Number of inputs of the system
  numStates = size(x_ir,1); % Number of states of the system
  numOutputs = size(e_ir,1); % Number of outputs of the system
  wsl = ws - wm; % slip speed (non persistent)

%-- Predictive parameters --%
  predHorizon = 3;
  ctrlHorizon = 2;

%-- Initialization of persistent --%
  if isempty(X_ir);
      X_ir = zeros(numStates,numCoeff);
  end
  if isempty(E_ir);
      E_ir = zeros(numOutputs*(numCoeff - 1),1);
  end
  if isempty(U_ir);
      U_ir = zeros(nInputs, numCoeff - 1 );
  end
  if isempty(Wsl);
      Wsl = ws;
  end
```

3. This step involves storing the past state values in the function of the number of coefficients of the polynomial, creation of the expanded state vector, as well as storage of the current and past errors due to the number of the polynomial coefficients.

```
%--------------- States of the system ---------------%
  X_ir = [x_ir, X_ir];
  X_ir = X_ir(:,1:numCoeff);
  X_xx_ir = Mult_DxMatrix(X_ir, D);

% System error vector in function of coefficients
  E_ir = [e_ir; E_ir];
  E_ir = E_ir(1:numOutputs*(numCoeff - 1));
```

4. The fourth step is the calculation of matrices, which are computed only if some change occurs in the state space model (if the slip speed ω_{sl} changes) or if it is the first calculation of the predictive matrix from Eq. (11.13). Within this conditional loop, repetitive matrices are also calculated, that is, those of the expanded model. Both calculations for predictive and repetitive matrices are performed with the help of functions that are explained later.

```
%----------- Computation of the matrices ------------%
  if ( (Wsl ~= wsl) || (isempty(A_pr_ir)) )
      Wsl = wsl;  % Assign a new value to the persistent slip speed

    %----------- Creating State Matrices ------------%
    %-- Filling matrix Am --%
      A_ad11 = (1-(Rr*Ts)/(sigma*Lr));
      A_ad_ir = [A_ad11, wsl*Ts; -wsl*Ts, A_ad11];
      C_cd_ir = eye(numOutputs);

    %-- Filling persistent matrix Bd --%
      if isempty(B_bd_ir);
          B_bd11 = Ts/(sigma*Lr);
          B_bd_ir = [B_bd11, 0; 0, B_bd11];
      end
    %------ Generating the repetitive matrices -------%
      A_ee_ir = A_ee_Matrix(A_ad_ir, C_cd_ir, D, numOutputs);
     B_ee_ir = B_ee_Matrix(C_cd_ir, B_bd_ir, numOutputs, numCoeff);

    %------ Generating the Predictive matrices -------%
     A_pr_ir = A_pr_Matrix(A_ee_ir, predHorizon);
     B_pr_ir = B_pr_Matrix(A_ee_ir, B_ee_ir, predHorizon, ctrlHorizon);
  end
```

5. It is now necessary to create the weighting matrices using the size of "Phi" and the creation of the `X` vector from Eq. (11.14), according to the theory of the controller and using the expanded state vector:

```
%------ Generating the weighting matrices -------%
  [m,n] = size(B_pr);
 Q_i_ir = 2e15*eye(m);
 R_v_ir = 3e5*eye(n);

%----------- Generating the future input -----------%
 X_ee_ir = [ X_xx_ir; E_ir];
```

6. As this controller does not have to generate more vectors, the minimized cost function is evaluated. Once we obtain the input vector (result of the cost function), the extended input vector is obtained and after that the immediate control signals are extracted using the function `current_value`, past inputs and the polynomial.

```
% Solving the minimization of the cost function.
 Vpr_ir = -(B_pr_ir'*Q_i_ir*B_pr_ir+R_v_ir)^-1*B_pr_ir'...
 ...*Q_i_ir*A_pr_ir*X_ee_ir;

% Extraction of the input variable
  U_xx = Vpr_ir(1:nInputs);
  [u_ir_0] = current_value(U_xx, U_ir, D); % immediate control signals
```

Besides the code of the controller itself, there are also the functions that are used to perform certain operations, such as the construction of the vectors to be used, expanded matrices, and predictive framework matrices, and extraction of values from the expanded vectors, such as prime examples.

During step 3, there is the `Mult_DxMatrix` function, which is the function to create the x_s vector for the expanded state vector. This function is

```
function [X_xx_ir] = Mult_DxMatrix( x_ir, D )
  % Initialization of variables
    numCoeff = size(D,2);
    numSamples = size(x_ir,2);
    X_xx_ir = zeros(size(x_ir,1), 1);
  % Verification
    if numCoeff ~= numSamples
        disp('wrong dimensions matrix, D != x');
        return
    end
  % Construction of Xs vector
    for i = 1: numSamples
        X_xx_ir = X_xx_ir + D(i)*x_ir(:,i);
    end
end
```

In step 4, there are four functions, two for the expanded matrices and two for the predictive matrices. The following two functions are those that help to create the expanded matrices for the expanded model:

```
function A_ee_ir = A_ee_Matrix( A_ad_ir, C_cd_ir, D, out_Size)
  % Initialization
    numstates = size(A_ad_ir,1);
    numCoeff = size(D,2);
    zeroA = zeros(numstates, (numCoeff - 1)*out_Size);
    I = eye(out_Size);
    zeroB = zeros((numCoeff - 2)*out_Size, numstates);
  % Construction of the matrix A
    C = [ C_cd_ir*A_ad_ir; zeroB];
    coder.varsize('A_pp', [1000 1000]);
    A_pp = -D(2)*I;
    for i = 3: numCoeff
        A_pp = [A_pp ,-D(i)*I];
    end
    Iaux = eye((numCoeff - 2)*out_Size, (numCoeff - 1)*out_Size);
    A_pp = [A_pp ; Iaux];
    A_ee_ir = [ A_ad_ir, zeroA; C, A_pp];
end

function [B_ee_ir] = B_ee_Matrix(C_cd_ir, B_bd_ir, out_Size, numCoeff)
  % Initialization
```

```
    numInputs = size(B_bd,2);
    zeroA = zeros((numCoeff - 2)*out_Size, numInputs);
  % Construction of the matrix B
    B_ee_ir = [ B_bd_ir; C_cd_ir*B_bd_ir; zeroA];
end
```

The other two functions are to construct the matrices of the predictive controller:

```
function [A_pr_ir] = A_pr_Matrix(A_ee_ir, predHorizon)
  % Initialization
    [row, col] = size(A_ee_ir);
    coder.varsize('A_pr_ir', [row*(predHorizon) col]);
  % Construction of the matrix
    A_pr_ir = A_ee_ir;
    Aaux = A_ee_ir;
    for i = 2: predHorizon
        Aaux = Aaux*A_ee_ir;
      A_pr_ir = [ A_pr_ir; Aaux];
    end
end

function [B_pr_ir] = B_pr_Matrix(A_ee_ir,...
...B_ee_ir, predHorizon, ctrlHorizon)
  % Initialization
    [rowSize, colSize] = size(B_ee_ir);
    zeroA = zeros(rowSize, colSize*(ctrlHorizon-1));
    B_prColumns = colSize*ctrlHorizon;
    coder.varsize('B_pr_ir', [rowSize*(predHorizon) B_prColumns]);
  % Construction of the matrix
    B_pr_ir = [ B_ee_ir, zeroA];
    Aux = B_pr_ir;
    for i = 1: (predHorizon - 1)
        Aux = [(A_ee_ir^(i))*B_ee_ir, Aux];
        Aux = Aux(:,1:B_prColumns);
      B_pr_ir = [ B_pr_ir; Aux];
    end
end
```

The last function is used to extract the immediate control signals from the expanded vector, using the past inputs, current expanded vector, and the polynomial.

```
function [x_ir] = current_value(X_xx_ir, X_ir, D )
  % Initialization
    numCoeff = size(D,2);
    numSamples = size(X_ir,2);
    x_ir = X_xx_ir;
```

```
  % Verification
    if (numCoeff - 1) ~= numSamples
        disp('wrong dimensions matrix, D != x');
        return
    end
  % extraction from the vector Us
    for i = 1: numSamples
        x_ir = x_ir - D(i+1)*X_ir(:,i);
    end
end
```

As the code of the standard model predictive control, this code can also be implemented experimentally on a DSP or other microprocessor, doing some optimizations and adjusting to the experimental board used. The implementation of this controller over the DFIG model is suggested as a useful tool for the reader.

11.4 Building simulation model for three-phase power CCG

The representation of the three-phase power CCG was presented in details in Chapter 2. In this way, the CCG in the $\alpha\beta$ frame (Section 2.3.2) can be described as follows:

$$v_{inv,\alpha} = R_g i_{g,\alpha} + L_g \frac{di_{g,\alpha}}{dt} + v_{g,\alpha} \tag{11.16}$$

and

$$v_{inv,\beta} = R_g i_{g,\beta} + L_g \frac{di_{g,\beta}}{dt} + v_{g,\beta}, \tag{11.17}$$

where $\vec{v}_{inv} = v_{inv,\alpha} + jv_{inv,\beta}$ expresses the inverter voltage vector, $\vec{i}_g = i_{g,\alpha} + ji_{g,\beta}$ represents the grid current vector, and $\vec{v}_g = v_{g,\alpha} + jv_{g,\beta}$ is the grid voltage vector. The variables R_g and L_g describe the internal resistance and inductance of the grid filter, respectively. Fig. 11.16 depicts how to apply the converter model in Simulink in three-phase systems. The elements in the $\alpha\beta$ frame can be computed using the abc-to-$\alpha\beta$ transformations for each measured vector, it is described in Section 2.1.1 and the three-phase model is stated in Section 2.3.1.

The active and reactive power can be calculated from the grid voltage and current vectors, according to Eqs. (11.18) and (11.19),

$$P_g = \frac{3}{2}\Re e[\vec{v}_{g,\alpha\beta} \cdot \vec{i}^{\,*}_{g,\alpha\beta}], \tag{11.18}$$

$$Q_g = \frac{3}{2}\Im m[\vec{v}_{g,\alpha\beta} \cdot \vec{i}^{\,*}_{g,\alpha\beta}]. \tag{11.19}$$

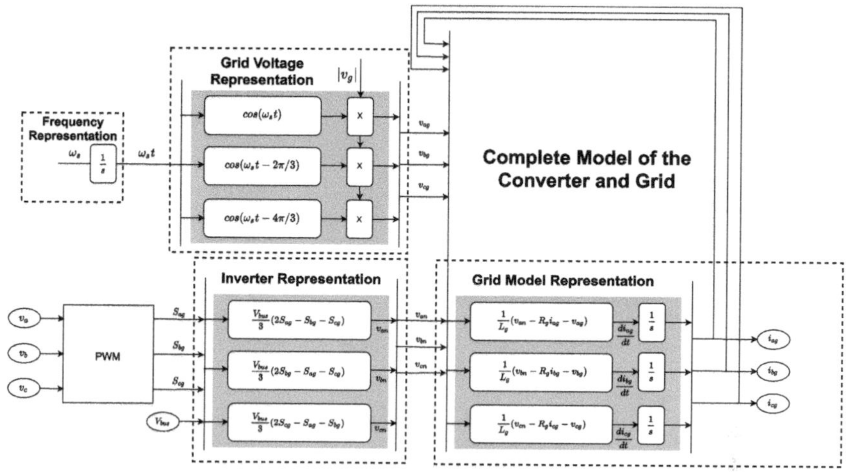

FIGURE 11.16 Complete three-phase power CCG model in Simulink.

11.4.1 Building simulation MPC for power converter

Applying Eqs. (11.16) and (11.17), it is possible to obtain the future values of the currents in the $\alpha\beta$ frame using the numerical derivative Euler's method and algebraic manipulation. Thus we obtain the equations for the future current values as follows:

$$i_{g,\alpha}(k+1) = \left(1 - \frac{R_g T}{L_g}\right) i_{g,\alpha}(k) + \frac{T}{L_g}(v_{inv,\alpha}(k) - v_{g,\alpha}(k)), \tag{11.20}$$

$$i_{g,\beta}(k+1) = \left(1 - \frac{R_g T}{L_g}\right) i_{g,\beta}(k) + \frac{T}{L_g}(v_{inv,\beta}(k) - \vec{v}_{g,\beta}(k)), \tag{11.21}$$

where T is the sampling time.

MPC is the strategy described in Section 7.3, and already applied in Section 11.3.1 to DFIG, so this subsection will approach the same technique but now using the converter representation to control the current in the $\alpha\beta$ frame.

For this controller, again the reader should use the output prediction equation given by

$$YY_g = AA_g x_g(k) + BB_g VV_g. \tag{11.22}$$

And the output minimized equation that contains the control actions is given by

$$VV_g = (BB_g^T W_{yy,g} BB_g + W_{uu,g})^{-1} BB_g^T W_{yy,g}(YY_{ref,g} - AAgx_g). \tag{11.23}$$

The function of the controller is to solve the latter equation (Eq. (11.23)), so in order to achieve this, all the matrices inside of it should be built.

The system to be controlled must be represented in the state space model, like the expression in (11.24) as introduced in Section 7.3:

$$\underbrace{\begin{bmatrix} i_{g,\alpha}(k+1) \\ i_{g,\beta}(k+1) \end{bmatrix}}_{x_g(k+1)} = \underbrace{\begin{bmatrix} 1-\dfrac{R_gT}{L_g} & 0 \\ 0 & 1-\dfrac{R_gT}{L_g} \end{bmatrix}}_{A_{ad,g}} \underbrace{\begin{bmatrix} i_{g,\alpha}(k) \\ i_{g,\beta}(k) \end{bmatrix}}_{x_g(k)} + \underbrace{\begin{bmatrix} \dfrac{T}{L_g} & 0 \\ 0 & \dfrac{T}{L_g} \end{bmatrix}}_{B_{bd,g}} \underbrace{\begin{bmatrix} v_{inv,\alpha}(k) - v_{g,\alpha}(k) \\ v_{inv,\beta}(k) - v_{g,\beta}(k) \end{bmatrix}}_{u_{u,g}(k)}. \tag{11.24}$$

The inductor used in series with grid and converter can be calculated as in (11.25), the resistance, which the reader should consider in series with the inductor, is its internal resistance,

$$L_g = \frac{V_{RMS}}{2\sqrt{6}f_0\Delta I}, \tag{11.25}$$

where V_{RMS} is the effective value of the voltage, f_0 is the switching frequency, and ΔI is the current variation in the inductor. For instance, the parameter can be described as:

```
%Grid Parameters
L=20e-3;%460/(2*sqrt(6)*12e3*0.25);
R=0.2;
```

Current references can be obtained from the active and reactive reference powers presented in (11.18) and (11.19), after algebraic manipulation and using the following equation:

$$\begin{bmatrix} i_{g,\alpha ref} \\ i_{g,\beta ref} \end{bmatrix} = \frac{2}{3(v_{g,\alpha}^2(k) + v_{g,\beta}^2(k))} \begin{bmatrix} v_{g,\alpha}(k) & v_{g,\beta}(k) \\ v_{g,\beta}(k) & -v_{g,\alpha}(k) \end{bmatrix} \begin{bmatrix} P_{gref} \\ Q_{gref} \end{bmatrix}. \tag{11.26}$$

And in Simulink the reader should do the matrix multiplication and build the block as it is depicted in Fig. 11.17.

The block in Fig. 11.18 can be summarized by listing inputs, outputs, and parameters:

Inputs. Grid currents $i_{\alpha\beta}$ and grid current references $i_{\alpha\beta Ref}$ in the stationary frame;

Outputs. Grid voltage in the stationary frame;

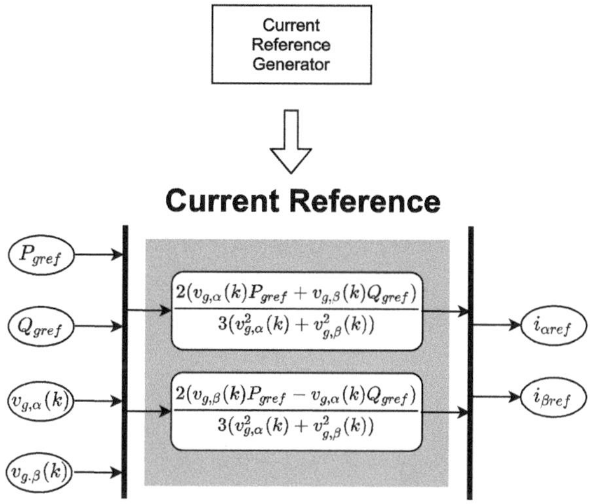

FIGURE 11.17 Current reference in Simulink.

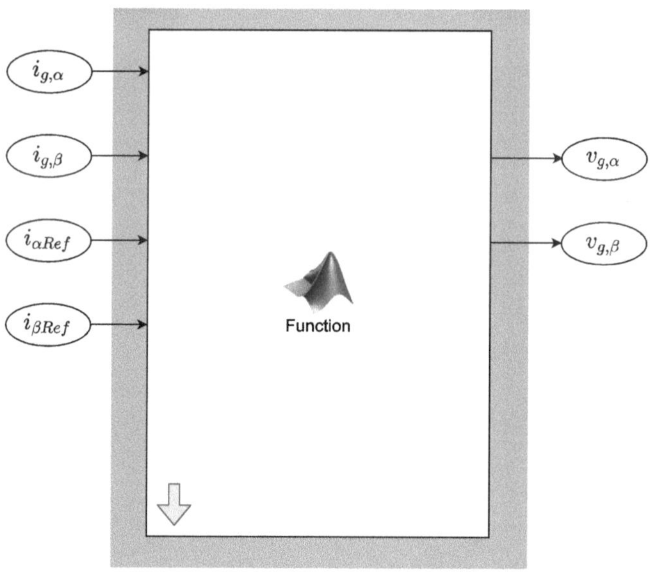

FIGURE 11.18 Model predictive control block, MATLAB/Simulink implementation.

Parameters. R_g, L_g, and T.

After choosing the "MATLAB function" block, setting the input, output, and parameters of the block, it is time to write the implementation code of the controller. The code will be explained step-by-step to facilitate reader's understanding:

1. The first lines of the code declare the control function, output, input, and parameters. The following lines declare the persistent variables across all the simulation with descriptive comments on the right-hand side:

```
function vab = PredRep(x_0, y_0, Ts, L, Rg)
%#codegen
coder.extrinsic('assignin')
coder.extrinsic('evalin')

  %---------- Declaring persistent variables ----------%

    persistent YYrefg;
    persistent Xx;
    persistent F_d;

    persistent Bm;

    persistent Fx;
    persistent Phi;
    persistent D;
    persistent Aux;
    persistent Aux2;
```

2. The second step is the declaration and initialization of normal variables, followed by initialization of persistent variables:

```
%--------- Declarations and Initializations ---------%
    nInputs = 2;
    numStates = size(x_0,1);
    numOutputs = size(y_0,1);
    numDisturbances = 2;

    %-- Predictive parameters --%
    predHorizon = 3;
    ctrlHorizon = 2;

    if isempty(YYrefg)
        YYrefg = zeros( predHorizon * numOutputs,1);
    end

    if isempty(Xx)
        Xx = zeros(numDisturbances, 1);
    end

    if isempty(F_d)
        F_d = 0;
    end
```

```
    if isempty(Aux)
        Aux = 0;
    end

    if isempty(Aux2)
        Aux2 = 0;
    end
```

3. The third step is the calculation of matrices, which are computed only if it is the first calculation of the predictive matrix from Eq. (11.22). The predictive control matrices are calculated with dedicated functions that will be explained later.

```
  %----------- Computation of the matrices ------------%

    if ((isempty(Fx)) )

      %----------- Obtaining State Matrices ------------%

      %-- Filling matrix Am --%

A11=(1-(Rg*T/L));

Am=[A11 0; 0 A11];

Bm=[T/Lg 0;0 T/Lg];

Cm=[1 0 ;0 1];

      %-- Obtaining the predictive matrices --%

        AAg = AA_ir_Matrix(Am, predHorizon, Cm);
        BBg = BB_ir_Matrix(Am, Bm, predHorizon, ctrlHorizon, Cm);
        D = 0 ;
    end
```

4. The fourth step is the creation of the weighting matrices using the size of the control matrix `BBg`.

```
  %------ Generating the weighting matrices ------%

    [m,n] = size(BBg);
                Wyyg = 2e20*eye(m);
                Wuug = 3e2*eye(n);
                assignin('base', 'AAg', AAg);
                assignin('base', 'BBg', BBg);
```

```
% This computation has to be done on EVERY step of the
% solver

    YYrefg = repmat(y_0, predHorizon, 1);
    Xx = repmat(wd, predHorizon, 1);

    U = ((BBg'*Wyyg*BBg + Wuug)^-1)*BBg'*Wyyg*(YYrefg -AAg*Xx);
    Kcoef = ((Phi'*Wy*Phi + Wu)^-1)*Phi'*Wy;
    assignin('base', 'Kcoef', Kcoef);

    u_0 = U(1:nInputs);
```

5. Now, the references vector is built repeating the current reference input (y_{ref}) with the prediction horizon. After this, it is time to do the most important step, which is the computation of the output minimized equation, Eq. (11.23).

```
% References and disturbance vectors
    YYrefg = repmat(y_0, predHorizon, 1);
    Xx = repmat(wd, predHorizon, 1);

% minimized cost function
    VVg = ((BBg'*Wyyg*BBg + Wuug)^-1)*BBg'*Wyyg*(YYrefg -AAg*xg);
    u_0 = VVg(1:nInputs);
```

The procedure for filling the matrices was already explained in Section 11.3.1

It is important to highlight that this code can also be implemented experimentally on some DSP or other experimental board, doing the needed modifications and simplifications. Finally, the authors encourage the readers to implement this controller on Simulink using the grid model which was already presented in the previous section.

11.5 Summary

The implementation of the DFIG and three-phase power CCG dynamic models, the rotor current control of the DFIG, and the current control of the CCG using the MPC-WM and rotor current control of the DFIG using the MPC-WM are described. In this way, the control schemes were implemented using MATLAB/Simulink language.

The block diagrams and the codes of the mentioned control techniques are introduced and they permit simulating the actions of the DFIG and CCG under several operation conditions.

Chapter 12

DFIG and three-phase power CCG experimental setup

Alfeu J. Sguarezi Filho
Federal University of ABC, Santo André, SP, Brazil

12.1 Experimental setups

This section introduces information about the hardware elements of the DFIG, three-phase power CCG, and PLL setups.

12.1.1 DFIG setup

The experimental results for the DFIG control in this book are obtained using electronic boards to measure the signals of the current and voltage, Texas Instruments model TMS320F28335 digital signal processor, as well as an encoder with 3600 pulses per revolution to acquire the speed and rotor position. The experimental setup is depicted in Fig. 12.1.

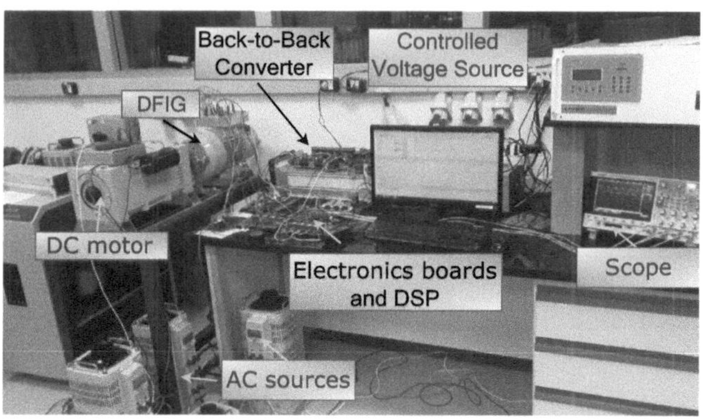

FIGURE 12.1 The experimental setup of DFIG used in MPC tests.

12.1.2 Three-phase CCG setup

The three-phase CCG setup has the main elements of the DFIG setup and it is depicted in Fig. 12.2. For this book results, we use electronic boards to measure

Model Predictive Control for Doubly-Fed Induction Generators and Three-Phase Power Converters
https://doi.org/10.1016/B978-0-32-390964-8.00021-X

the signals of the current and voltage, Texas Instruments model TMS320F28335 digital signal processor, and inductive filters $L = 20$ mH; and the main parameters are described in Table B.1.

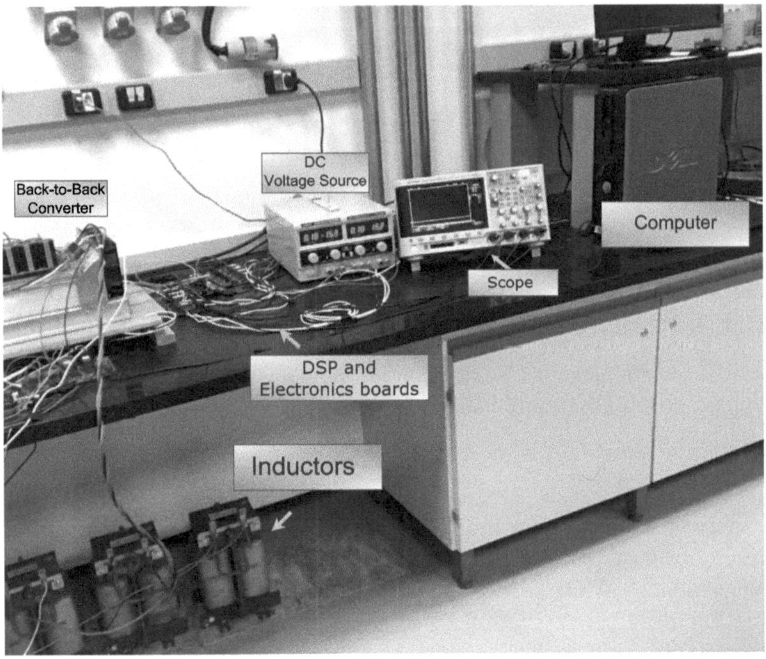

FIGURE 12.2 The experimental setup of three-phase CCG used in MPC tests.

12.1.3 The PLL setup

The PLL setup is depicted in Fig. 12.3. This setup has the main elements of the DFIG/Three-phase CCG setup. For this book results, we used Texas Instruments model TMS320F28335 digital signal processor, electronic boards to measure the signals of the current and voltage, and a resistive load bank. This setup permitted testing several PLLs algorithms before their implementation in the DFIG or three-phase CCG control.

12.1.4 Data acquisition, power supply, and DC motor

The components described in this section were employed to the setup for DFIG, three-phase CCG, and PLL. The electronic boards are designed at the LEPS and they are made using electronics component in accordance with the levels of the voltage and current of the DFIG and the input voltage that the DSP operates, namely from 0 to 3.3 V. In this case, the levels of the DFIG are presented in

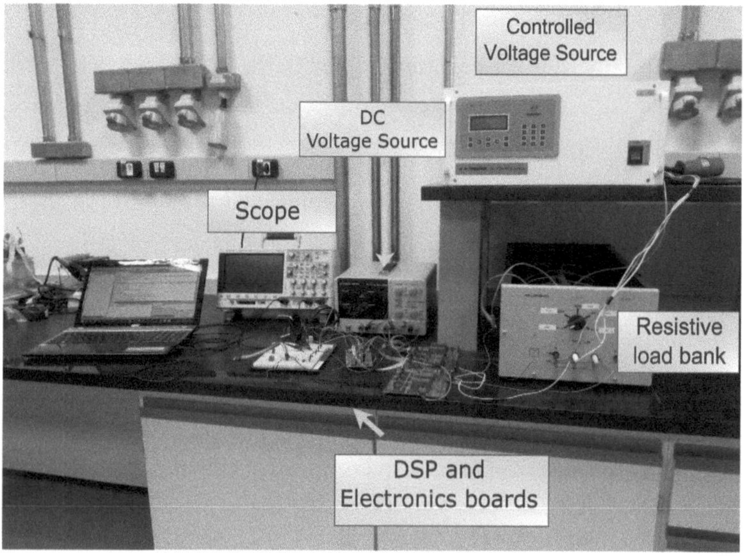

FIGURE 12.3 The experimental setup for PLL tests.

Table A.1, and the manufacturer is Delorenzo. The current and voltage sensors are from LEM LV20-P and LA55-P, and some information about the electronics board design can be seen in [130,131]. The electronics boards can be constructed by using basic electronic components. Basically, the measured voltages and current are transformed into voltage signals between 0 and 3.3 V. The boards are fed by a DC/DC voltage source ±15 V and 5 V.

The tests under distorted voltage were performed using a controllable electronic voltage source, Supplier model FCATHQ 450-38-50-n55210. The three-phase converter connected to the rotor of DFIG was a two-level voltage source converter Semikron model SKS20F(B6CI)2P + E1CIF + B6U14V 12. This is a back-to-back configuration, which has 12 IGBTs and a three-phase uncontrolled rectifier. The IGBTs are driven using 15 V/0 V voltage levels.

One converter of the mentioned configuration was employed when the setup was for the three-phase power CCG. The DC bus voltage could be made by a three-phase uncontrolled rectifier connected to the electrical grid or by a controllable electronic DC voltage source. It can be mentioned that one possibility is to use just transformers connected to the mains directly when the test is under normal operation conditions for DFIG or for three-phase power CCG. In this way, the transformers permit tuning voltage levels.

A 3 kW, 220 V independent excited DC motor drives the rotor speed of the DFIG. This motor is fed by means of the three-phase uncontrolled rectifier connected to the electrical grid, and the control is in open loop by means of the armature voltage control due to the low dynamics in the DFIG applications, such as wind. The manufacturer is Delorenzo.

12.1.5 System initialization

This section describes the procedures adopted to the DFIG and CCG setup initialization.

12.1.5.1 System initialization using DFIG

Before start, the stator of the DFIG is not connected to the grid in the power system and there is no voltage in the DC bus of the DFIG setup. There is a contactor between the grid and the stator of the DFIG. And the setup can start using the following steps:

- Charge the DC link bus of the back-to-back converter – it can be made by an uncontrolled rectifier or by a CCG, or by a DC voltage source;
- Start the control algorithm – it provides the regulation of the voltage amplitude of the stator by means of the rotor current control;
- Start the grid synchronization method – it permits the regulation of the voltage angle of the stator;
- Connect the stator of DFIG to the grid by turning on the contactor – it allows DFIG operating as a generator;
- Make the desired tests.

Regarding the grid synchronization method, it is important to highlight that the initial rotor position is not the same as the electrical grid. In this way, it is necessary to perform the synchronization of the rotor and grid angles. In this way, the work [132] developed a method to accomplish this task. In this case, a comparison is made between the direct component of the stator voltage vector and the same component of the grid voltage vector, and a PI controller processes this error, generating the increment of θ_r to perform the grid synchronization. Another PI controller processes the error between the quadrature components of the stator voltage vector and the grid voltage vector; and the target is to reach the grid voltage peak value. In this case, the PI controller calculates a reference of the direct component of the rotor current vector. Hence, when the stator voltages have the same peak value and phase of the grid voltages, the contactor can be turned on. It is important to highlight that the electrical grid position is computed by the PLL and, when the DFIG links the mains, the stator and grid voltages are the same.

12.1.5.2 System initialization using three-phase CCG

Before start, the terminals of the three-phase power CCG are not connected to the grid in the power system in the setup initialization. There is a contactor between them. And, it can start using the fallow steps:

- Charge the DC link bus – this can be made by an uncontrolled rectifier or by a DC voltage source;
- Start the control algorithm – it provides the regulation of the voltage amplitude of the converter by means of the current control;

- Start the grid synchronization method – it permits the regulation of the voltage angle of the grid by using the PLL;
- Connect the terminals of the CCG to the grid by turning on the contactor – it allows the three-phase CCG to operate injecting current into the electrical grid;
- Make the desired tests.

Regarding the grid synchronization method, it is important to highlight that the electrical grid position is computed by the PLL. Another PI controller processes the error between the direct components of the converter and grid voltage vectors. In this context, the target is to reach the grid voltage peak value. In this case, the PI controller calculates a reference of the quadrature component of the grid current vector. Hence, when the terminals of the three-phase power converter CCG voltages have the same peak value and phase of the grid voltages, the contactor can be turned on.

12.2 Information about the microcontroller

As mentioned in the last section, the microcontroller is a DSP TMS320F28335 [133]. This DSP has the following features:

- 16 analog digital converters (ADC) with 12-bit resolution
- 18 PWM channels
- 2 modules to process quadrature encoder signals
- Hardware implementation for SVM
- 150 MHz clock
- Clock and system control
- 64 general purpose pins which can be connected to one of the eight external core interrupts

The DFIG and three-phase CCG experimental setups use just two interrupts, one to process the measured signals and the other to the MPC, transformations, and PWM algorithm. The first can employ the frequency greater than the MPC frequency. However, it can be the same. This is up to the designer. In the DFIG experimental setup, the MPC operates at 10 kHz, and in the three-phase CCG experimental setup, the MPC operates at 20 kHz. The DSP can be programmed in C, C++, or assembly languages.

12.2.1 Functionality of the microcontroller

The DPS has several ways to program its mains functions. In this context, the TMS320F28335 was programmed using the ControlSUITE for C2000 microcontrollers that allows minimizing the system development due to its documentation, tools, and set of software with examples. And, it has some features as:

- Centralized repository for C2000
- Graphical user interface (GUI) for intuitive navigation of software, libraries, development kits, application notes, and user guides

- Device software and support
- Kit software and support

12.3 Predictive control implementation

There are basically two MPC techniques in this book, namely FCS-MPC and MPC/MPRC. These techniques were programmed using the DSP. The algorithms are presented to each technique in the corresponding chapter. Regarding the implementation, initially, the FCS-MPC has a somewhat more complex implementation compared to MPC/MPRC due to the computation of matrices in DSP of the MPC/MPRC and when FCS-MPC has lower prediction horizon (n_y). However, when the n_y increases, the computational cost of the FCS-MPC increases exponentially because the controller has to test all states of the converter during the computation of the control algorithm. In the case of the MPC/MPRC, the size of the matrices also increases. Hence, in this book we choose a low value of the prediction horizon.

On the other hand, it is possible to calculate the voltage vector using the minimized cost function of the MPC/MPRC. Thus, it is possible to employ high value control and prediction horizons, choose the weighting matrices gain by using simulations. Also, this approach is valid when it has constraints. So, one has to program a control law when the constraints are detected. However, other problems such as incorrect output prediction or problems as mentioned in the respective control chapters will appear. In this case, the predictive control will lose performance or it can become unstable.

The general pseudocode for the MPC implementation in this book is presented in Algorithm 12.1.

12.3.1 Building embedded controller in DSP

This section introduces some tips to the basic code to embed the controller in DSP. One is a microprocessor chip with specific characteristics, it is an electrical architecture optimized to operate with digital signal processing. This section shows how the reader can embed the controller presented in Chapter 11 due to the fact the DSP needs various files to operate. The idea is to give basic information, and the ControlSUITE by Texas Instruments can help in this task.

It is necessary to load all the MACRO codes responsible to make the:

- interruptions
- measurements of the signals such as current, speed, angle, voltages, etc.
- estimates of the fluxes, power or current, etc.
- transformations
- modulation techniques

1. First, the code in the C language for voltage and current Clarke transformation is presented below:

Algorithm 12.1 General MPC algorithm

```
while MPC active do
    if System measurements for DFIG then
        Stator and rotor currents and voltages, rotor speed and position
        Coordinates transformations, PLL activation, stator flux magnitude estimation
    end if
    if System measurements for three-phase CCG then
        Grid and converter voltages, grid current
        Coordinates transformations, PLL activation
    end if
    Load the MPC algorithm: MPF-FCS or MPC/MPRC-WM
    if MPC/MPRC - WM then
        Calculate the voltage vector components of the rotor (DFIG) or converter (CCG) cases, respectively
        Coordinate transformations, if it is necessary
        Apply the voltage vector calculated by the MPC using PWM technique
    end if
    if MPF-FCS then
        Apply the state of the converter that the least absolute error
    end if
end while
```

```
volt.Alpha = _IQmpy((volt.A -_IQmpy(volt.B, _IQ(0.5))
    -_IQmpy(volt.C,_IQ(0.5))), _IQ(0.666666667));
volt.Beta = _IQmpy(_IQmpy(volt.B, _IQ(0.866)) -_IQmpy(volt.C,
    _IQ(0.866)), _IQ(0.666666667));

current.Alpha = _IQmpy((current.A -_IQmpy(current.B, _IQ(0.5))
    -_IQmpy(current.C,_IQ(0.5))), _IQ(0.666666667));
current.Beta = _IQmpy(_IQmpy(current.B, _IQ(0.866)) -_IQmpy(current.C,
    _IQ(0.866)), _IQ(0.666666667));
```

2. As the $\alpha\beta$ values are between 0 and 1, now they are scaled to their real value by multiplying with their nominal value:

```
VAlpha = _IQmpy(volt.Valpha,_IQ(VA));
VBeta = _IQmpy(volt.Vbeta,_IQ(VB));

IAlpha = _IQmpy(current.Alpha,_IQ(IA));
IBeta = _IQmpy(current.Beta,_IQ(IB));
```

3. As the third step, now the modules for voltage and currents are obtained:

```
ModV = sqrt(VAlpha*VAlpha + VBeta*VBeta)*_IQ(1/sqrt(2));
ModI = sqrt(IAlpha*IAlpha + IBeta*IBeta)*_IQ(1/sqrt(2));
```

4. The voltage references are obtained in the function of the voltage module and the PLL angle `anguloVg`:

```
VAlphaRef = _IQcosPU(anguloVg)*ModV*_IQ(sqrt(2));
VBetaRef = _IQsinPU(anguloVg)*ModV*_IQ(sqrt(2));
```

5. Then with `VAlphaRef`, `VBetaRef`, and the equation presented in Section 11.4.1, it is possible to obtain the references for the current alpha and beta.

```
Kref = 2/(3*(VAlphaRef*VAlphaRef + VBetaRef*VBetaRef)) ;
IAlpha_ref = Kref*(VAlphaRef*P_ref + VBetaRef*Q_ref);
IBeta_ref = Kref*(VBetaRef*P_ref-VAlphaRef*Q_ref);
```

6. Also in Section 11.4.1 we presented an equation to obtain the control law, which can be embedded to DSP as a gain; in this example code, the gain is obtained through Simulink/MATLAB®.

7. Now that the output of the controller was calculated (voltage in the $\alpha\beta$ frame), it is necessary to give it as input to the system through the commutation. The `svgen1` variable is assigned the corresponding output values scaled to be from 0 to 1; right after that, the call of the macro in line 3 calculates the appropriate duty ratios to generate the given reference voltage using the space vector PWM technique. Once the times are obtained, they are assigned to the `pwm1` variable to be used in the commutation.

```
1  svgen1.Ualpha = UAlpha_out/(160); // Input of the alpha element
2  svgen1.Ubeta = UBeta_out/(160); // Input of the beta element
3  SVGENDQ_MACRO(svgen1)
4
5  pwm1.MfuncC1 = svgen1.Ta;
6  pwm1.MfuncC2 = svgen1.Tb;
7  pwm1.MfuncC3 = svgen1.Tc;
8  PWM_MACRO(1,2,3,pwm1)
```

8. Start again.

12.4 Summary

This chapter described the experimental setups built at the laboratory to make the tests using the DFIG, CCG, and PLL. The main elements of each setup, the basic algorithm, and the basic information of the setups were introduced. Additionally, information about the DSP employed and the start procedures of the DFIG and CCG were presented.

Appendix A

DFIG parameters

The IM parameters are presented in Table A.1.

TABLE A.1 IM parameters.

Parameter	Value
Stator resistance per phase (R_s)	1 Ω
Stator inductance per phase (L_s)	0.2010 H
Rotor resistance per phase (R_r)	3.122 Ω
Rotor inductance per phase (L_r)	0.2010 H
Mutual inductance (L_m)	0.1917 H
Stator nominal frequency	60 Hz
Pole pairs (NP)	2
Nominal active power (P_s)	3 kW
Nominal rotor voltage (V_r)	440 Y V
Nominal stator voltage (V_s)	220/380 Δ−Y V
Sampling frequency	10 kHz
Switching frequency (f_{sw})	10 kHz

Appendix B

Three-phase power CCG parameters

The three-phase power CCG parameters are presented in Table B.1.

TABLE B.1 Three-phase power CCG parameters.

Parameter	Value
Inductor internal resistance per phase (R_g)	0.10 Ω
Inductance per phase (L_g)	22 mH
Nominal grid frequency	60 Hz
Nominal grid voltage (V_g)	220 V
Sampling frequency	20 kHz
Switching frequency (f_{sw})	20 kHz

Appendix C

DC link voltage representation

The converters of the grid and rotor side can be connected by means of the DC link. In this way, the energy stocked in the capacitor permits maintaining the voltage at a constant level in which this is a component of the back-to-back converter. The mentioned voltage can be expressed by means of the capacitors' current (i_{c-dc}) and it can be described by

$$v_{dc} = \frac{1}{C}\int i_{c-dc}dt. \tag{C.1}$$

The current of the capacitor can be expressed as per the current from the resistance in parallel with the capacitor i_{dc}, from the current of the grid side converter (i_{g-dc}), and from the current of the rotor side converter (i_{r-dc}), as follows:

$$\begin{aligned}
i_{c-dc} &= i_{r-dc} - i_{g-dc} - i_{dc},\\
i_{g-dc} &= S_{ag}i_{g,a} + S_{bg}i_{g,b} + S_{cg}i_{g,c},\\
i_{r-dc} &= -S_{ar}i_{r,a} - S_{br}i_{r,b} - S_{cg}i_{r,c},\\
i_{dc} &= \frac{v_{dc}}{R}.
\end{aligned} \tag{C.2}$$

Each DC current element i_{g-dc} and i_{r-dc} from the converters in Eq. (C.2) are represented by means of the current in three-phase systems ($i_{g,abc}$ and $i_{r,abc}$) and the states of the semiconductors of each converter (S_{abcr} and S_{abcg}).

Bibliography

[1] B.K. Bose, Power Electronics and Motor Drives: Advances and Trends, Elsevier Science, Amsterdam, the Netherlands, 2020.

[2] V. Yaramasu, B. Wu, Model Predictive Control of Wind Energy Conversion Systems, IEEE Press Series on Power Engineering, Wiley, 2016, https://books.google.com.br/books?id=PTlxDQAAQBAJ.

[3] G. Abad, J. Lopez, M. Rodríguez, L. Marroyo, G. Iwanski, Doubly-Fed Induction Machine: Modeling and Control for Wind Energy Generation, Vol. 85, John Wiley & Sons, 2011.

[4] J.A. Rossiter, Model Based Predictive Control - a Practical Approach, CRC Press, 2003.

[5] J. Rodriguez, P. Cortes, Predictive Control of Power Converters and Electrical Drives, Wiley - IEEE, Wiley, 2012, https://books.google.com.br/books?id=I2JgX5ghSq4C.

[6] P.K. Kovács, E. Rácz, Transient Phenomena in Electrical Machines, Elsevier, Amsterdam, the Netherlands, 1984.

[7] E. Clarke, Circuit Analysis of AC Power Systems, John Wiley & Sons, Inc., New York, 1950.

[8] R.H. Park, Two reaction theory of synchronous machines. generalized method of analysis – part I, in: Proceedings of the Winter Convention of the AIEE, 1929, pp. 716–730.

[9] A.E. Fitzgerald, S.D.U.C. Kingsley, Electric Machinery, McGraw-Hill, 2003.

[10] M. Rashid, Power Electronics Handbook, Academic Press Series in Engineering, Academic Press, 2001, https://books.google.com.br/books?id=072rwRgFpcgC.

[11] A. Kwasinski, P.T. Krein, P.L. Chapman, Time domain comparison of pulse-width modulation schemes, IEEE Power Electronics Letters 1 (3) (2003) 64–68.

[12] Keliang Zhou, Danwei Wang, Relationship between space-vector modulation and three-phase carrier-based pwm: a comprehensive analysis [three-phase inverters], IEEE Transactions on Industrial Electronics 49 (1) (2002) 186–196.

[13] Texas Instruments Incorporated, Implementation of a speed field oriented control of three phase AC Induction Motor using TMS320F240, 1998.

[14] V. Toro, Electric Machines and Power Systems, Prentice-Hall, 1985, https://books.google.com.br/books?id=Y-RSAAAAMAAJ.

[15] W. Leonhard, Control of Electrical Drives, Springer-Verlag Berlin Heidelberg New York Tokyo, 1985.

[16] C. Ong, Dynamic Simulation of Electric Machinery: Using MATLAB/SIMULINK, Prentice Hall PTR, 1998, https://books.google.com.br/books?id=_OweAQAAIAAJ.

[17] S. Yamamura, Spiral Vector Theory of AC Circuits and Machines, Clarendon Press, Oxford, 1992.

[18] P. Sen, Principles of Electric Machines and Power Electronics, third edition, Wiley Global Education, 2013, https://books.google.com.br/books?id=X35bAgAAQBAJ.

[19] R. Krishnan, Electric Motor Drives: Modeling, Analysis, and Control, Prentice Hall, 2001, https://books.google.com.br/books?id=1lgtAQAAMAAJ.

[20] M.L. de Aguiar, M.M. Cad, The concept of complex transfer functions applied to the modeling of induction motors, in: 2000 IEEE Power Engineering Society Winter Meeting. Conference Proceedings (Cat. No. 00CH37077), Vol. 1, 2000, pp. 387–391.

[21] J. Holtz, The dynamic representation of ac drive systems by complex signal flow graphs, in: Proceedings of 1994 IEEE International Symposium on Industrial Electronics (ISIE'94), 1994, pp. 1–6.

[22] P. Krause, O. Wasynczuk, S. Sudhoff, I.P.E. Society, Analysis of Electric Machinery and Drive Systems, IEEE Press Series on Power Engineering, Wiley, 2002, https://books.google.com.br/books?id=8m4vAQAAIAAJ.

[23] S. Chapman, Electric Machinery Fundamentals, McGraw-Hill Series in Electrical and Computer Engineering, McGraw-Hill Companies, Incorporated, 2005, https://books.google.com.br/books?id=XGn2pQsQaN4C.

[24] J. Holtz, L. Springob, Reduced harmonics pwm controlled line-side converter for electric drives, IEEE Transactions on Industry Applications 29 (4) (1993) 814–819.

[25] N. Mohan, T. Undeland, W. Robbins, Power Electronics: Converters, Applications, and Design, Power Electronics: Converters, Applications, and Design, John Wiley & Sons, 2003, https://books.google.com.br/books?id=oxR8vB2XjgIC.

[26] B. Bose, Power Electronics in Renewable Energy Systems and Smart Grid: Technology and Applications, IEEE Press Series on Power Engineering, Wiley, 2019, https://books.google.com.br/books?id=qAyfDwAAQBAJ.

[27] D. Holmes, T. Lipo, Pulse Width Modulation for Power Converters: Principles and Practice, IEEE Press Series on Power Engineering, Wiley, 2003, https://books.google.com.br/books?id=8LGi1AjSfpcC.

[28] S. Mishra, S.K. Gagrai, M. Singh, An ann based simulation and control of grid integrated dfig for 2 mw small hydro power plant, in: 2019 6th International Conference on Signal Processing and Integrated Networks (SPIN), 2019, pp. 540–546.

[29] P. Qian, B. Feng, H. Liu, X. Tian, Y. Si, D. Zhang, Review on configuration and control methods of tidal current turbines, Renewable & Sustainable Energy Reviews 108 (2019) 125–139, https://doi.org/10.1016/j.rser.2019.03.051, http://www.sciencedirect.com/science/article/pii/S1364032119301911.

[30] A. Petersson, L. Harnefors, T. Thiringer, Comparison between stator-flux and grid-flux-oriented rotor current control of doubly-fed induction generators, in: 2004 IEEE 35th Annual Power Electronics Specialists Conference (IEEE Cat. No. 04CH37551), Vol. 1, 2004, pp. 482–486.

[31] A. Luiz Lacerda Ferreira Murari, J. Alberto Torrico Altuna, R. Vani Jacomini, C. Mario Rocha Osorio, J. Sebastian Solis Chaves, A.J. Sguarezi Filho, A proposal of project of pi controller gains used on the control of doubly-fed induction generators, IEEE Latin American Transactions 15 (2) (2017) 173–180.

[32] G.F. Franklin, J.D. Powel, M.l. Workman, Digital Control of Dynamic Systems, Addison-Wesley Publishing Company, 1994.

[33] A.J. Sguarezi Filho, E. Ruppert, A deadbeat active and reactive power control for doubly fed induction generator, Electric Power Components and Systems 38 (5) (2010) 592–602, https://doi.org/10.1080/15325000903376966.

[34] C. Rocha-Osorio, J. Solís-Chaves, L.L. Rodrigues, J.A. Puma, A. Sguarezi Filho, Deadbeat–fuzzy controller for the power control of a doubly fed induction generator based wind power system, ISA Transactions 88 (2019) 258–267, https://doi.org/10.1016/j.isatra.2018.11.038, https://www.sciencedirect.com/science/article/pii/S0019057818304804.

[35] R. Franco, C.E. Capovilla, R.V. Jacomini, J.A.T. Altana, A.J. Sguarezi Filho, A deadbeat direct power control applied to doubly-fed induction aerogenerator under normal and sag voltages conditions, in: IECON 2014 - 40th Annual Conference of the, IEEE Industrial Electronics Society, IEEE, 2014.

[36] J. Solis-Chaves, M.S. Barreto, M.B. Salles, V.M. Lira, R.V. Jacomini, A.J. Sguarezi Filho, A direct power control for dfig under a three phase symmetrical voltage sag condition, Control Engineering Practice 65 (2017) 48–58, https://doi.org/10.1016/j.conengprac.2017.05.002, http://www.sciencedirect.com/science/article/pii/S0967066117301107.

[37] R. Teodorescu, M. Liserre, P. Rodrígues, Grid Converter for Photovoltaic and Wind Power Systems, John Wiley, United Kingdom, 2011.
[38] U.P. Yagnik, M.D. Solanki, Comparison of l, lc lcl filter for grid connected converter, in: 2017 International Conference on Trends in Electronics and Informatics (ICEI), 2017, pp. 455–458.
[39] R.N. Beres, X. Wang, M. Liserre, F. Blaabjerg, C.L. Bak, A review of passive power filters for three-phase grid-connected voltage-source converters, IEEE Journal of Emerging and Selected Topics in Power Electronics 4 (1) (2016) 54–69, https://doi.org/10.1109/JESTPE.2015.2507203.
[40] A.M. Hava, U. Ayhan, V.V. Aban, A dc bus capacitor design method for various inverter applications, in: 2012 IEEE Energy Conversion Congress and Exposition (ECCE), 2012, pp. 4592–4599.
[41] M. Rashid, Power Electronics Circuits, Devices and Applications, Prentice Hall, 2004.
[42] A. Yazdani, R. Iravani, Voltage-Sourced Converters in Power Systems: Modeling, Control, and Applications, Wiley - IEEE, Wiley, 2010, https://books.google.com.br/books?id=_x_4Cu-BKwkC.
[43] R. Teodorescu, M. Liserre, P. Rodriguez, Grid Converters for Photovoltaic and Wind Power Systems, Wiley - IEEE, Wiley, 2011, https://books.google.com.br/books?id=5k4tmcVIaM0C.
[44] D. Novotny, T. Lipo, Vector Control and Dynamics of AC Drives, Monographs in Electrical and Electronic Engineering, Clarendon Press, 1996, https://books.google.com.br/books?id=5FAHb_z4eAQC.
[45] P. Vas, Sensorless Vector and Direct Torque Control, Monographs in Electrical and Electronic Engineering, Oxford University Press, 1998, https://books.google.com.br/books?id=9ORSAAAAMAAJ.
[46] F. Briz, M.W. Degener, R.D. Lorenz, Analysis and design of current regulators using complex vectors, IEEE Transactions on Industry Applications 32 (2000) 817–825.
[47] E.F. Camacho, C.B. Alba, Model Predictive Control, 2nd edition, Springer Science & Business Media, 2000.
[48] J. Rodriguez, P. Cortes, Predictive Control of Power Converters and Electrical Drives, Vol. 40, John Wiley & Sons, 2012.
[49] J. Holtz, S. Stadtfeld, A predictive controller for the stator current vector of ac machines fed from a switched voltage source, in: International Power Electronics Conference, IPEC, 1983, pp. 1665–1675.
[50] L. Zhang, R. Norman, W. Shepherd, Long-range predictive control of current regulated pwm for induction motor drives using the synchronous reference frame, IEEE Transactions on Control Systems Technology 5 (1) (1997) 119–126.
[51] R. Kennel, A. Linder, M. Linke, Generalized predictive control (gpc)-ready for use in drive applications?, in: IEEE Power Electronics Specialists Conference - PESC 4, 2001, pp. 1839–1844.
[52] K.S. Low, A digital control technique for a single-phase pwm inverter, IEEE Transactions on Industrial Electronics 45 (4) (1998) 672–674.
[53] H. Abu-Rub, J. Guzinski, Z. Krzeminski, H.A. Toliyat, Predictive current control of voltage-source inverters, IEEE Transactions on Industrial Electronics 51 (3) (2004) 585–593.
[54] Z. Xin-fang, X. Da-ping, L. Yi-bing, Predictive functional control of a doubly fed induction generator for variable speed wind turbines, in: IEEE World Congress on Intelligent Control and Automation, June 2004.
[55] K. Drobnic, M. Nemec, D. Nedeljkovic, V. Ambrozic, Predictive direct control applied to ac drives and active power filter, IEEE Transactions on Industrial Electronics 56 (6) (2009) 1884–1893.
[56] J. Rodriguez, M.P. Kazmierkowski, J.R. Espinoza, P. Zanchetta, H. Abu-Rub, H.A. Young, C.A. Rojas, State of the art of finite control set model predictive control in power electronics, IEEE Transactions on Industrial Informatics 9 (2) (2013) 1003–1016.

[57] A. Dehghanzadeh, G. Farahani, H. Vahedi, K. Al-Haddad, Model predictive control design for dc-dc converters applied to a photovoltaic system, International Journal of Electrical Power & Energy Systems 103 (2018) 537–544, https://doi.org/10.1016/j.ijepes.2018.05.004, http://www.sciencedirect.com/science/article/pii/S0142061517330016.

[58] B. Cao, B.M. Grainger, X. Wang, Y. Zou, G.F. Reed, Z.H. Mao, Direct torque model predictive control of a five-phase permanent magnet synchronous motor, IEEE Transactions on Power Electronics 36 (2) (2021) 2346–2360.

[59] P. Karamanakos, T. Geyer, Guidelines for the design of finite control set model predictive controllers, IEEE Transactions on Power Electronics 35 (7) (2020) 7434–7450.

[60] C. Zheng, T. Dragičević, B. Majmunović, F. Blaabjerg, Constrained modulated model-predictive control of an lc-filtered voltage-source converter, IEEE Transactions on Power Electronics 35 (2) (2020) 1967–1977.

[61] M. Bermúdez, C. Martín, F. Barrero, X. Kestelyn, Predictive controller considering electrical constraints: a case example for five-phase induction machines, IET Electric Power Applications 13 (8) (2019) 1079–1088.

[62] M. Abdelrahem, R. Kennel, Efficient direct model predictive control for doubly-fed induction generators, Electric Power Components and Systems 45 (5) (2017) 574–587, https://doi.org/10.1080/15325008.2017.1289572.

[63] L. Tarisciotti, P. Zanchetta, A. Watson, J. Clare, M. Degano, S. Bifaretti, Modulated model predictive control (m2pc) for a 3-phase active front-end, in: 2013 IEEE Energy Conversion Congress and Exposition, 2013, pp. 1062–1069.

[64] J. Hu, Z.Q. Zhu, Improved voltage-vector sequences on dead-beat predictive direct power control of reversible three-phase grid-connected voltage-source converters, IEEE Transactions on Power Electronics 28 (1) (2013) 254–267.

[65] L. Tarisciotti, P. Zanchetta, A. Watson, J.C. Clare, M. Degano, S. Bifaretti, Modulated model predictive control for a three-phase active rectifier, IEEE Transactions on Industry Applications 51 (2) (2015) 1610–1620.

[66] M. Rivera, F. Morales, C. Baier, J. Muñoz, L. Tarisciotti, P. Zanchetta, P. Wheeler, A modulated model predictive control scheme for a two-level voltage source inverter, in: 2015 IEEE International Conference on Industrial Technology (ICIT), 2015, pp. 2224–2229.

[67] E.S. De Santana, E. Bim, W.C. do Amaral, A predictive algorithm for controlling speed and rotor flux of induction motor, IEEE Transactions on Industrial Electronics 55 (12) (2008) 4398–4407, https://doi.org/10.1109/tie.2008.2007376.

[68] J.M. Maciejowski, Predictive Control: With Constraints, Pearson Education, 2002.

[69] A. Linder, R. Kanchan, P. Stolze, R. Kennel, Model-Based Predictive Control of Electric Drives, Cuvillier Verlag, 2010, https://books.google.com.br/books?id=Udv-DwAAQBAJ.

[70] L. Wang, S. Chai, D. Yoo, L. Gan, K. Ng, PID and Predictive Control of Electrical Drives and Power Converters Using MATLAB / Simulink, Wiley - IEEE, Wiley, 2015, https://books.google.com.br/books?id=IVk0CgAAQBAJ.

[71] T. Geyer, Model Predictive Control of High Power Converters and Industrial Drives, Wiley, 2016, https://books.google.com.br/books?id=ovVmCgAAQBAJ.

[72] L.L. Rodrigues, O.A. Vilcanqui, E.R. Conde D., J.M. Guerero, A.J. Sguarezi Filho, Generalized predictive control applied to the dfig power control using state-space model and voltage constraints, Electric Power Systems Research 182 (2020) 106227, https://doi.org/10.1016/j.epsr.2020.106227, https://www.sciencedirect.com/science/article/pii/S0378779620300341.

[73] T. Hadjina, I. Mišković, M. Baotić, Finite control set model predictive control of a grid connected two-level converter, in: 2017 19th International Conference on Electrical Drives and Power Electronics (EDPE), 2017, pp. 353–358.

[74] K. Wróbel, P. Serkies, K. Szabat, Model predictive base direct speed control of induction motor drive—continuous and finite set approaches, Energies 13 (5) (2020) 1193.

[75] R.V. Jacomini, A.J. Sguarezi Filho, Finite control set applied to the direct power control of a dfig operating under voltage sags, IEEE Transactions on Sustainable Energy 10 (2) (2019) 952–960, https://doi.org/10.1109/TSTE.2018.2856463.

[76] D. Casadei, G. Serra, K. Tani, Implementation of a direct control algorithm for induction motors based on discrete space vector modulation, IEEE Transactions on Power Electronics 15 (4) (2000) 769–777, https://doi.org/10.1109/63.849048.
[77] M. Amiri, J. Milimonfared, D.A. Khaburi, Predictive torque control implementation for induction motors based on discrete space vector modulation, IEEE Transactions on Industrial Electronics 65 (9) (2018) 6881–6889, https://doi.org/10.1109/TIE.2018.2795589.
[78] X. Fang, S. Misra, G. Xue, D. Yang, Smart grid—the new and improved power grid: a survey, IEEE Communications Surveys and Tutorials 14 (4) (2012) 944–980.
[79] M. Hashmi, S. Hänninen, K. Mäki, Survey of smart grid concepts, architectures, and technological demonstrations worldwide, in: Innovative Smart Grid Technologies (ISGT Latin America), 2011 IEEE PES Conference on, IEEE, 2011, pp. 1–7.
[80] M. Fadaeenejad, A.M. Saberian, M. Fadaee, M. Radzi, H. Hizam, M. AbKadir, The present and future of smart power grid in developing countries, Renewable & Sustainable Energy Reviews 29 (2014) 828–834.
[81] S. Borlase, Smart Grids: Infrastructure, Technology, and Solutions, CRC Press, 2017.
[82] P. Kulshrestha, L. Wang, M.-Y. Chow, S. Lukic, Intelligent energy management system simulator for phevs at municipal parking deck in a smart grid environment, in: 2009 IEEE Power & Energy Society General Meeting, IEEE, 2009, pp. 1–6.
[83] M. Pourbehzadi, T. Niknam, J. Aghaei, G. Mokryani, M. Shafie-khah, J.P. Catalao, Optimal operation of hybrid AC/DC microgrids under uncertainty of renewable energy resources: a comprehensive review, International Journal of Electrical Power & Energy Systems 109 (2019) 139–159.
[84] Z. Ozkan, A.M. Hava, Three-phase inverter topologies for grid-connected photovoltaic systems, in: International Power Electronics Conference, 2014, pp. 498–505.
[85] Q. Shen, B. Ramachandran, S.K. Srivastava, M. Andrus, D.A. Cartes, Power and energy management in integrated power system, in: IEEE Electric Ship Technologies Symposium, 2011, pp. 414–419.
[86] R.M. Hudson, M.R. Behnke, R. West, S. Gonzalez, J. Ginn, Design considerations for three-phase grid connected photovoltaic inverters, in: IEEE Photovoltaic Specialists Conference, 2002, pp. 1396–1401.
[87] L. Xu, D. Zhi, L. Yao, Direct power control of grid connected voltage source converters, in: IEEE Power Engineering Society General Meeting, 2007, pp. 1–6.
[88] R. Smolenski, J. Bojarski, A. Kempski, P. Lezynski, Time-domain-based assessment of data transmission error probability in smart grids with electromagnetic interference, IEEE Transactions on Industrial Electronics 61 (4) (2014) 1882–1890.
[89] T. Sauter, M. Lobashov, End-to-end communication architecture for smartgrids, IEEE Transactions on Industrial Electronics 58 (4) (2011) 1218–1228.
[90] J.G. Proakis, Digital Communications, McGraw-Hill, New York, 2008.
[91] S. Lin, D.J. Costello, Error Control Coding, Pearson Education India, 2004.
[92] H. Rohling, OFDM Concepts for Future Communication Systems, Springer-Verlag Berlin Heidelberg, 2011.
[93] H.-L. Hung, Adaptive resource allocation algorithm based on minimize average bit-error-rate for OFDM systems, Wireless Personal Communications 94 (4) (2017) 3091–3112.
[94] A. Lunardi, A. Sguarezi Filho, C. Capovilla, I. Casella, A. de Medeiros, A wireless coded predictive direct power control for renewable energy sources in smart grid environment, International Journal of Electrical Power & Energy Systems 112 (2019) 319–325, https://doi.org/10.1016/j.ijepes.2019.05.004, http://www.sciencedirect.com/science/article/pii/S014206151832324X.
[95] R. Gallager, Low Density Parity Check Codes, Research Monograph Series, 1963.
[96] Y. Zhang, W.E. Ryan, Toward low ldpc-code floors: a case study, IEEE Transactions on Communications 57 (6) (2009) 1566–1573.
[97] H. Jin, A. Khandekar, R. McEliece, Irregular repeat-accumulate codes, in: Proc. 2nd Int. Symp. Turbo Codes and Related Topics, Citeseer, 2000, pp. 1–8.

[98] Y. Zhang, W.E. Ryan, Y. Li, Structured eira codes with low floors, in: Proceedings. International Symposium on Information Theory, 2005, ISIT 2005, IEEE, 2005, pp. 174–178.

[99] T.J. Richardson, R.L. Urbanke, The capacity of low-density parity-check codes under message-passing decoding, IEEE Transactions on Information Theory 47 (2) (2001) 599–618, https://doi.org/10.1109/18.910577.

[100] L. Dinoi, F. Sottile, S. Benedetto, Design of versatile eira codes for parallel decoders, IEEE Transactions on Communications 56 (12) (2008) 2060–2070, https://doi.org/10.1109/TCOMM.2008.050022.

[101] C.E. Capovilla, I.R.S. Casella, A.J.S. Filho, T.A. dos Santos Barros, E.R. Filho, Performance of a direct power control system using coded wireless OFDM power reference transmissions for switched reluctance aerogenerators in a smart grid scenario, IEEE Transactions on Industrial Electronics 62 (1) (2015) 52–61.

[102] C.K. Duffey, R.P. Stratford, Update of harmonic standard IEEE-519: IEEE recommended practices and requirements for harmonic control in electric power systems, IEEE Transactions on Industry Applications 25 (6) (1989) 1025–1034, https://doi.org/10.1109/28.44238.

[103] A.J. Sguarezi Filho, M.E. de Oliveira Filho, E. Ruppert Filho, A predictive power control for wind energy, IEEE Transactions on Sustainable Energy 2 (1) (2011) 97–105.

[104] J. Holtz, J. Quan, J. Pontt, J. Rodriguez, P. newman, H. Miranda, Design of fast and robust current regulators for high-power drives based on complex state variables, IEEE Transactions on Industry Applications 40 (2004) 1388–1397.

[105] A. Sguarezi Filho, E. Filho, Model-based predictive control applied to the doubly-fed induction generator direct power control, IEEE Transactions on Sustainable Energy 3 (3) (July 2012) 398–406, https://doi.org/10.1109/TSTE.2012.2186834.

[106] A. Lunardi, E. Conde, J. Assis, D. Fernandes, A. Sguarezi Filho, Model predictive control with modulator applied to grid inverter under voltage distorted, Energies 14 (16) (2021) 4953, https://doi.org/10.3390/en14164953.

[107] H. Akagi, Y. Kanazawa, A. Nabae, Instantaneous reactive power compensators comprising switching devices without energy storage components, IEEE Transactions on Industry Applications IA-20 (3) (1984) 625–630, https://doi.org/10.1109/TIA.1984.4504460.

[108] S. Hara, Y. Yamamoto, T. Omata, M. Nakano, Repetitive control system: a new type servo system for periodic exogenous signals, IEEE Transactions on Automatic Control 33 (7) (1988) 659–668, https://doi.org/10.1109/9.1274.

[109] K. Zhou, F. Blaabjerg, D. Wang, Y. Yang, Periodic Control of Power Electronic Converters, Energy Engineering, Institution of Engineering and Technology, 2016, https://books.google.com.br/books?id=loCHDQAAQBAJ.

[110] R. Costa-Castello, J. Nebot, R. Grino, Demonstration of the internal model principle by digital repetitive control of an educational laboratory plant, IEEE Transactions on Education 48 (1) (2005) 73–80, https://doi.org/10.1109/TE.2004.832873.

[111] L. Wang, From Plant Data to Process Control: Ideas for Process Identification and PID Design, Systems and Control Series, Taylor & Francis, 2000, https://books.google.com.br/books?id=aFEVAAAAIAAJ.

[112] L. Wang, S. Chai, E. Rogers, C.T. Freeman, Multivariable repetitive-predictive controllers using frequency decomposition, IEEE Transactions on Control Systems Technology 20 (6) (2012) 1597–1604.

[113] L. Wang, A tutorial on predictive repetitive control, in: 2016 Australian Control Conference (AuCC), IEEE, 2016, pp. 114–119.

[114] E.R. Conde D., A. Lunardi, A.J. Sguarezi Filho, Current control for dfig systems under distorted voltage using predictive repetitive control, IEEE Journal of Emerging and Selected Topics in Power Electronics 9 (4) (2021) 4354–4363, https://doi.org/10.1109/JESTPE.2020.3025503.

[115] D. Xu, F. Blaabjerg, W. Chen, N. Zhu, Advanced Control of Doubly Fed Induction Generator for Wind Power Systems, IEEE Press Series on Power Engineering, Wiley, 2018, https://books.google.com.br/books?id=B4liDwAAQBAJ.

[116] Ieee recommended practices and requirements for harmonic control in electrical power systems, in: IEEE Std 519-1992, 1993, pp. 1–112.
[117] M.M. Islam, Power Quality: Harmonics, 2018, pp. 158–184.
[118] J. Hu, H. Nian, H. Xu, Y. He, Dynamic modeling and improved control of dfig under distorted grid voltage conditions, IEEE Transactions on Energy Conversion 26 (1) (2011) 163–175, https://doi.org/10.1109/TEC.2010.2071875.
[119] J. Hu, H. Xu, Y. He, Coordinated control of dfig's rsc and gsc under generalized unbalanced and distorted grid voltage conditions, IEEE Transactions on Industrial Electronics 60 (7) (2013) 2808–2819, https://doi.org/10.1109/TIE.2012.2217718.
[120] J. Das, Power System Analysis: Short-Circuit Load Flow and Harmonics, second edition, Power Engineering (Willis), CRC Press, 2017, https://books.google.com.br/books?id=tDLNBQAAQBAJ.
[121] D. Wolaver, Phase-Locked Loop Circuit Design, Prentice Hall Biophysics and B, Prentice Hall, 1991, https://books.google.com.br/books?id=C2V5QgAACAAJ.
[122] K. Ogata, Modern Control Engineering, Instrumentation and Controls Series, Prentice Hall, 2010, https://books.google.com.br/books?id=Wu5GpNAelzkC.
[123] V. Kaura, V. Blasko, Operation of a phase locked loop system under distorted utility conditions, IEEE Transactions on Industry Applications 33 (1) (1997) 58–63, https://doi.org/10.1109/28.567077.
[124] S. Chung, Phase-locked loop for grid-connected three-phase power conversion systems, IEE Proceedings. Electric Power Applications 147 (3) (2000) 213–219, https://doi.org/10.1049/ip-epa:20000328.
[125] P. Rodriguez, A. Luna, M. Ciobotaru, R. Teodorescu, F. Blaabjerg, Advanced grid synchronization system for power converters under unbalanced and distorted operating conditions, in: IECON 2006 - 32nd Annual Conference on IEEE Industrial Electronics, 2006, pp. 5173–5178.
[126] S. Golestan, M. Ramezani, J.M. Guerrero, F.D. Freijedo, M. Monfared, Moving average filter based phase-locked loops: performance analysis and design guidelines, IEEE Transactions on Power Electronics 29 (6) (2014) 2750–2763, https://doi.org/10.1109/TPEL.2013.2273461.
[127] M. Abdelrahem, C. Hackl, R. Kennel, Finite position set-phase lockedloop for sensorless control of direct-driven permanent-magnet synchronous generators, IEEE Transactions on Industry Applications 33 (4) (2008) 3097–3105, https://doi.org/10.1109/TPEL.2017.2705245.
[128] M. Abdelrahem, C.M. Hackl, R. Kennel, J. Rodríguez, Computationally efficient finite-position-set-phase-locked loop for sensorless control of pmsgs in wind turbine applications, IEEE Transactions on Power Electronics 36 (3) (2021) 3007–3016, https://doi.org/10.1109/TPEL.2020.3015575.
[129] F. Lino, J. Assis, D.A. Fernandes, R. Jacomini, F.F. Costa, A.J. Sguarezi Filho, One-cycle Fourier finite position set pll, Energies 14 (7) (2021), https://doi.org/10.3390/en14071824, https://www.mdpi.com/1996-1073/14/7/1824.
[130] A. Sedra, K. Smith, Microeletrônica, Pearson Makron Books, 2000.
[131] Simplifying current sensing - how to design with current sense amplifiers, Report, Texas Instruments, 2020, https://www.ti.com/lit/eb/slyy154a/slyy154a.pdf?ts=1604247386647.
[132] A.G. Abo-Khalil, D. Lee, S. Lee, Grid connection of doubly-fed induction generators in wind energy conversion system, in: 2006 CES/IEEE 5th International Power Electronics and Motion Control Conference, Vol. 3, 2006, pp. 1–5.
[133] Tms320f2833x, tms320f2823x digital signal controllers (dscs), Report, Texas Instruments, 2010, https://www.ti.com/lit/ds/symlink/tms320f28335.pdf?HQS=TI-null-null-digikeymode-df-pf-null-wwe&ts=1604264345781.

Index

A

Actions control, 43, 44, 50, 103, 121, 133, 173, 185
Additive white Gaussian noise (AWGN), 75
Analog digital converter (ADC), 195
Angle
  grid, 149, 153, 155, 194
  rotor, 171
  voltage, 161, 162, 194, 195
Armature voltage control, 193

B

Base vectors, 21–23
Bidirectional converter, 1
Bus voltage, 21, 162, 193

C

Clarke transformation (CT), 157, 158, 196
Components
  CCG, 138
  electronic, 193
  grid voltage, 152
  grid voltage vector, 142
  harmonic, 4, 19, 111, 117, 119–122, 147–150, 153, 155
  rotor current vector, 87, 89, 91, 128–130
  rotor voltage, 34
  rotor voltage vector, 35, 57, 87, 94
  stator voltage, 29
  vector, 101, 125, 136, 140, 163
  voltage vector, 114
Constituent elements, 35, 36
Control
  actions, 43, 44, 50, 103, 121, 133, 173, 185
  algorithm, 1, 194, 196
  bits, 73
  deadbeat, 37
  design, 123, 134
  DFIG, 4, 191
  effort, 44, 47, 50, 51
  flexibility, 68
  function, 162, 175, 188
  grid current, 39, 40, 99, 100, 105, 138, 139, 143
  horizon, 50
  law, 43, 196, 198
  loops, 32, 108
  matrix, 176, 189
  model predictive, 3, 4, 43, 173
  output, 51
  power, 3, 35, 69
  predictive, 2, 3, 45, 177, 184, 196
  repetitive, 107, 115
  rotor current, 3, 30, 31, 40, 87, 91, 128–130, 190, 194
  rotor current vector, 85
  samples, 113, 125
  schemes, 190
  signal, 44, 52, 181, 183
  strategy, 33, 178
  system loop, 107
  techniques, 2, 41, 43, 45, 190
  theory, 33
  vector, 28, 37, 40
Controllable electronic
  DC voltage source, 193
  voltage source, 193

Controller
effort, 143
formulation, 52
implementations, 157
performance, 108
repetitive, 111
robustness, 77
Converter
electronic, 27
power, 2, 53, 77, 185
power components, 83
representation, 185
switches, 2
voltage, 16, 133
voltage vector, 17, 38, 40, 99
Converter connected to the grid (CCG), 14–16, 28, 31, 37, 38, 51, 53, 99, 105, 106, 114
components, 138
current control, 37, 101, 134
current control applications, 106
current control design, 134
current control diagram, 40
fundamental frequency, 137
model, 39, 132, 167
representation, 16, 17
space state, 98
three-phase, 97, 135, 136, 191, 192
three-phase power, 69, 71, 114, 122, 131, 134, 138, 184, 185, 190, 191, 193
Coordinating transformation, 10, 12
Cost function, 2, 3, 43–48, 50, 70, 71, 87, 94, 100, 104, 113, 125, 137, 153, 178, 181
Cyclic prefix (CP), 74, 75

D

DC
bus voltage, 193
converters, 45
link, 1, 14, 21, 27, 28, 31, 38, 122, 134, 167
link voltage, 40
motor, 63, 89, 95, 128, 192, 193
signals, 108, 148
voltage source, 193, 194
Deadbeat
control, 37
for DFIG, 37
technique, 34
theory, 40
controllers, 3, 37
rotor current control for DFIG, 33
Descriptive comments, 162, 175, 179, 188
Digital signal processor (DSP), 76
Direct power control (DPC), 35, 36
concepts, 40
deadbead strategy, 37
representation, 37
technique, 36
Discrete Fourier transform (DFT), 109
Discrete representation, 33, 40, 57, 66, 93, 106, 125, 136, 143
Distorted
grid voltage, 123, 133, 134, 142
stator voltage, 120
voltage, 4, 117–119, 133, 134, 139, 142, 155, 193
operation, 129, 141
peak value, 129
representation, 155
Disturbance
in electrical grid, 147
matrix, 48, 49
prediction, 87, 94
vector, 176–178
Doubly-fed induction generator (DFIG), 1, 2, 27, 31, 33, 36, 40, 45, 51, 53, 57, 58, 63, 65, 85, 86, 88, 91, 93, 94, 114, 118, 123–125, 167, 168, 171, 173, 178, 191, 192
applications, 193
control, 4, 191
current control, 87
deadbeat rotor current control, 35
DPC, 91, 95
experimental setup, 195
links, 194
mathematical model, 169
model, 118, 119, 123, 169, 170, 178, 184
nominal power, 27

operation, 121
power, 35
representation, 55
rotor, 27, 28, 51, 88, 123, 193
rotor current control, 85, 123, 127
setup, 191, 194
speed operation, 31
stator, 30, 144, 194
Dual second order generalized integrator (DSOGI), 149, 150, 153, 155
Duty cycle, 47, 48, 58, 59, 61, 62, 71

E

Electrical grid, 2–4, 14, 15, 27, 28, 38, 40, 51, 67–69, 82, 100, 136, 139, 193–195
Electrical grid voltage, 30, 117, 147
Electromagnetic torque, 11, 13, 25, 28, 169, 171
Electronic
components, 193
converter, 27
power converters, 1
Elements
grid current, 101
grid voltage vector, 138, 148
harmonic, 144, 149, 155
rotor current vector, 85, 126, 128
rotor voltage, 37, 57
rotor voltage vector, 37, 85, 92, 128
stator flux, 56
Error signal, 114, 126, 137

F

Fast Fourier transform (FFT), 74, 75, 80
Filter
elements, 38
grid, 184
Finite
angle size, 153
control set, 3, 45
control set model predictive control, 45, 47
position set, 149, 152
Finite impulse response (FIR), 151
Fluxes vectors, 11, 13
Forward error correction (FEC), 68
Fundamental
component, 19, 20, 120–122, 133
concepts, 3, 4, 55, 155
frequency, 110, 114, 117, 119, 123, 126, 129, 133, 135, 147, 148, 153
frequency component, 138
frequency voltage, 117
harmonic component, 18

G

Graphical user interface (GUI), 195
Grid
active, 39, 133
active power, 101, 139
angle, 149, 153, 155, 194
angle positions, 152
current, 16, 69, 76, 78, 100–102, 117, 134, 138, 141, 142, 186
control, 39, 40, 99, 100, 105, 138, 139, 143
elements, 101
signal, 77
vector components, 100
distorted voltage, 131
distorted voltage peak value, 141
disturbances, 30, 149
filter, 184
frequency, 149, 153
impedance, 15
operation, 30
power, 3, 83, 102, 140, 156
power control, 101, 139
reactive, 97, 101, 138
reactive power, 101, 139
side converter, 39, 40, 134
voltage, 4, 16, 30, 39, 71, 78, 102, 117, 133, 135, 138, 149, 184, 186, 194, 195
components, 152
distortions, 132, 139, 144
elements, 148
peak value, 194, 195
vector, 17, 30, 39, 40, 101, 134, 135, 140, 148, 152, 153, 184, 194, 195

vector components, 100, 142
vector elements, 138, 148
Grid voltage orientated control (GVOC), 39, 40

H

Harmonic
components, 4, 19, 111, 117, 119–122, 147–150, 153, 155
elements, 144, 149, 155

I

Imaginary component, 11, 12, 14, 121, 134
Impulse filter, 108
Induction machine (IM), 3, 5, 7, 9
Inductor filter, 16
Inputs prediction, 87, 94, 100
Insulated gate bipolar transistor (IGBT), 14
Intelligent controllers, 2
Internal model principle (IMP), 107, 108
Inverse Clarke transformation, 159
Inverse fast Fourier transform (IFFT), 74
Inverter
bottom switches, 167
output, 18
switches, 21
top switches, 167
voltage vector, 184

L

Loop filter (LF), 145
Low density parity check (LDPC), 68
coding, 68, 71, 78
coding schemes, 75, 83

M

MACRO codes, 196
Matrices predictive control, 176, 189
Metal oxide semiconductor field effect transistor (MOSFET), 14
Microcontroller, 195
Minimized cost function, 2, 43, 45, 52, 58, 88, 95, 101, 111, 113, 115, 125, 137, 152, 181, 196
Model predictive
controllers, 3
repetitive controllers, 4
Model predictive control (MPC), 3, 4, 43, 173, 195
Model predictive repetitive control (MPRC), 109, 111, 113–115, 117, 123, 196
for DFIG, 178
fundamentals, 109
system, 114
technique, 111
Modulation
space vector, 21, 70, 160
techniques, 2, 18
Moving average filter (MAF), 149, 151, 153
Multipath components, 75

N

Null vectors, 21, 22, 48, 59

O

Operation
DFIG, 121
distorted voltage, 129, 141
frequency, 75
grid, 30
Orthogonal frequency division multiplexing (OFDM), 68, 74
Outputs
components prediction, 94
prediction, 87, 100
Overmodulation region, 19

P

Peak value, 5, 18, 147, 167, 194, 195
distorted voltage, 129
grid voltage, 194, 195
Performance
controller, 108
rotor current, 89
Periodic signal, 107, 111
Periodic signal representation, 109
Persistent variables, 162, 175, 179, 188
Phase detector (PD), 145

Phase-locked loop (PLL), 30, 31, 39, 40, 192, 194, 198
PI controller, 31, 32, 39, 40, 152, 194, 195
Power
  control, 3, 35, 69
  control for DFIG, 3, 35
  converter, 2, 53, 77, 185
  DFIG, 35
  electronic converter, 2, 15, 27
  electronic converters, 25, 40, 68
  electronic drives, 45
  electronics, 1, 2, 4, 5, 45, 47, 49, 160
  grid, 3, 83, 102, 140, 156
  losses, 122, 134
  model, 67
  predictive control, 69
  reactive, 31, 36, 40, 65, 77, 83, 94, 95, 102, 134, 141, 142, 184
  stator, 31, 36, 37, 63, 95
  system, 68, 194
Prediction
  disturbance, 87, 94
  effort weighting matrices, 51
  horizon, 47, 50, 176, 190, 196
  model, 50
Predictive
  control, 2, 3, 45, 177, 184, 196
    matrices, 176, 189
    method, 70
    system, 71
    techniques, 2
  controller, 2, 44, 76, 111, 183
  direct control, 45
  matrix, 176, 180, 189
Pulse width modulation (PWM), 18, 160
  algorithm, 31, 40, 51, 114, 138, 195
  signals, 21–23
  sinusoidal, 18, 19
  space vector, 95, 101
  technique, 18, 35, 37, 88, 126

Q

Quadrature
  axis, 29, 30, 33, 89
  axis component, 40
  component, 147, 194, 195
  encoder signals, 195
Quadrature signal generator (QSG), 149
Quaternary phase shift keying (QPSK), 73
  symbol, 74
  symbol subvectors, 74

R

Reactive
  grid, 97, 101, 138
  power, 31, 36, 40, 65, 77, 83, 94, 95, 102, 134, 141, 142, 184
  power representation, 119, 133
  stator, 34–37, 56, 85, 88, 90, 92–95, 120, 128
  stator power, 65
Renewable energy source (RES), 68, 71
Repetitive
  controller, 111
  predictive controller, 4
Repetitive control (RC), 107, 115
  actions, 143
  fundamentals, 4, 107
Representation
  CCG, 15–17
  converter, 185
  DFIG, 55
  distorted voltage, 155
  rotor voltage, 55
  space vector, 5, 25
Resonant controller, 110
Rotor
  active, 12, 14
  active power, 122
  angle, 171
  angular frequency, 9
  angular speed, 179
  current, 4, 29, 34, 65, 88, 95, 121, 126, 128, 171, 174, 179
    control, 3, 30, 31, 40, 87, 91, 128–130, 190, 194
    control for DFIG, 85, 105, 126, 143
    error, 33
    in rotor frame, 91
    performance, 89
    vector, 31, 33, 35, 55, 87–90, 118, 123, 128, 129, 194
    vector components, 87, 89, 91, 128–130
    vector control, 85, 86
    vector elements, 85, 126, 128
  DFIG, 27, 28, 51, 88, 123, 193

dynamics, 85, 92
equations, 125
flux, 35, 36, 171
flux vector, 36
frame, 31, 95
leakage inductance, 10
mechanical speed, 171
model, 13, 118
position, 191, 194
resistance, 9, 36, 37
side, 36
side converter, 31, 36, 121
side converter control, 1
space state representation, 85, 92
vector, 34
voltage, 9, 37, 57, 65, 119, 123, 174, 179
components, 34
elements, 37, 57
representation, 55
vector, 10, 12, 31–34, 36, 56–58, 63, 89, 93, 95, 118, 121
vector components, 33, 35, 57, 87, 94
vector elements, 37, 85, 92, 128

S

Sampling
instant, 44, 57, 111, 113, 123, 125, 135, 136
time, 34, 45, 48, 51, 52, 86, 88, 93, 95, 99, 100, 108, 114, 115, 126, 137, 138, 151, 153, 185
Settling time, 32, 51, 63, 89, 95, 101, 104, 128, 139, 155
Signal
control, 44, 52, 181, 183
generator, 111, 149
grid current, 77
Silicon controlled rectifier (SCR), 14
Simulink
block, 157
environment, 174
software, 173, 178
Sinusoidal
input signal, 150
pulse width modulation, 19
PWM, 18, 19
Sinusoidal PWM (SPWM), 18, 25
Slip angle, 31
Smart grid (SG), 4, 67
central control, 71
environment, 3
infrastructure, 68
operator, 69, 75, 76
system, 80
Space state, 111
Space state equations, 85, 98
Space vector, 5, 7, 9, 10, 16–18, 169
notation, 5
PWM, 95, 101
PWM technique, 198
representation, 5, 25
stator flux, 28
Space vector modulation (SVM), 21, 70, 160
State
space, 2, 86, 93, 99, 111–114, 124–126, 136, 173, 175, 178, 180, 186
space representation, 111, 112, 135, 137
vector, 181
Stationary frame, 16, 31, 170, 186
Stator
active, 12, 14, 30, 31, 36, 94, 95, 119
active power, 29, 34, 63, 65, 95
apparent power, 63, 95
apparent power elements, 65
current, 33, 63, 65, 89, 90, 95, 118, 120, 126, 128, 129
current vector, 13, 55, 65, 95
current vector elements, 123
DFIG, 30, 144, 194
distorted voltage, 118
distorted voltage operation, 129
flux, 29, 30, 33, 87, 94, 174, 176
elements, 56
orientation, 29, 30
space vector, 28
vector, 28, 30, 118
vectors elements, 118
leakage inductance, 10
power, 31, 36, 37, 63, 95
power components control, 97
power elements, 95

reactive, 34–37, 56, 85, 88, 90, 92–95, 120, 128
reactive power, 29, 34, 63, 65, 95
resistance, 9, 29, 30, 118
voltage, 30, 31, 56, 57, 65, 89, 95, 118–121, 123, 126, 194
components, 29
space vector, 29
vector, 10, 12, 30, 194
Stator voltage oriented control (SVOC), 30, 31, 36, 55, 57, 123
Step tests, 63, 89, 95, 128
Switches
converter, 2
inverter, 21

T

THSPWM, 19, 20, 25
Total harmonic distortion (THD), 117, 118, 129, 130
Triangular signal, 160

U

Unbalance input voltage vector, 150
Uncontrolled rectifier, 194
Universal serial bus (USB), 76

V

Vector
components, 101, 125, 136, 140, 163
control, 28, 37, 40
fundamental concepts, 30
fundamentals, 38
technique, 37
disturbance, 176, 178
fundamental concepts control for DFIG, 3
grid voltage, 17, 30, 39, 40, 101, 134, 135, 140, 148, 152, 153, 184, 194, 195
rotor, 34
rotor current, 31, 33, 35, 55, 87–90, 118, 123, 128, 129, 194
rotor voltage, 10, 12, 31–34, 36, 56–58, 63, 89, 93, 95, 118, 121
sector, 22
stator flux, 28, 30, 118
stator voltage, 10, 12, 30, 194
voltage, 2, 21, 30, 33, 39, 48, 51, 70, 71, 88, 114, 119, 126, 138, 147, 196
Voltage
amplitude, 194
angle, 161, 162, 194, 195
converter, 16
distorted, 4, 117–119, 133, 134, 139, 142, 155, 193
distortions, 117, 119
distortions representation, 117
grid, 4, 16, 30, 39, 71, 78, 102, 117, 133, 135, 138, 149, 184, 186, 194, 195
levels, 193
rotor, 9, 37, 57, 65, 119, 123, 174, 179
rotor vector, 58
sag, 30, 56, 148, 149, 153, 155
signals, 19, 193
source, 15, 117, 193
source inverter, 45
stator, 30, 31, 56, 57, 65, 89, 95, 118–121, 123, 126, 194
value, 18, 21, 47
vector, 2, 21, 30, 33, 39, 48, 51, 70, 71, 88, 114, 119, 126, 138, 147, 196
Voltage-controlled oscillator (VCO), 145

W

Weighting matrices, 51, 95, 101, 103, 106, 176, 181, 189
Weighting matrices gain, 196
Wireless control system, 77
Wireless power control system, 83

Z

Zeroth order hold (ZOH), 174, 179

FSC
www.fsc.org
MIX
Papier | Fördert gute Waldnutzung
FSC® C083411

Kolibri 360 GmbH
Ferdinand-Jühlke-Straße 7
99095 Erfurt, Deutschland
produktsicherheit@kolibri360.de